Becoming Yellow

BECOMING YELLOW

A Short History of Racial Thinking

Michael Keevak

PRINCETON UNIVERSITY PRESS
PRINCETON AND OXFORD

Copyright © 2011 by Princeton University Press
Published by Princeton University Press,
41 William Street, Princeton, New Jersey 08540

In the United Kingdom:
Princeton University Press,
6 Oxford Street, Woodstock,
Oxfordshire OX20 1TW
press.princeton.edu

All Rights Reserved

Library of Congress Cataloging-in-Publication Data

Keevak, Michael, 1962–
  Becoming yellow : a short history of racial thinking / Michael Keevak.
    p. cm.
  Includes bibliographical references and index.
  ISBN 978-0-691-14031-5 (hardcover : acid-free paper) 1. Racism—Western countries—History—18th century. 2. Racism—Western countries—History—19th century. 3. Race awareness—Western countries—History—18th century. 4. Race awareness—Western countries—History—19th century. 5. East Asians—Race identity. 6. National characteristics, East Asian. I. Title.
  HT1523.K44 2011
  305.8009182′109033—dc22

                                                                2010046654

British Library Cataloging-in-Publication Data is available

Support generously provided by the Chiang Ching-Kuo Foundation
for International Scholarly Exchange

This book has been composed in Minion with Clarendon display
Printed on acid-free paper. ∞
Printed in the United States of America
10 9 8 7 6 5 4 3 2 1

Contents

List of Illustrations   vii

Acknowledgments   ix

Introduction
No Longer White: The Nineteenth-Century
Invention of Yellowness   1

Chapter 1
Before They Were Yellow: East Asians in Early Travel
and Missionary Reports   23

Chapter 2
Taxonomies of Yellow: Linnaeus, Blumenbach, and the
Making of a "Mongolian" Race in the Eighteenth Century   43

Chapter 3
Nineteenth-Century Anthropology and the Measurement
of "Mongolian" Skin Color   70

Chapter 4
East Asian Bodies in Nineteenth-Century Medicine:
The Mongolian Eye, the Mongolian Spot, and "Mongolism"   101

Chapter 5
Yellow Peril: The Threat of a "Mongolian" Far East,
1895–1920   124

Notes   145

Works Cited   175

Index   211

Illustrations

COLOR PLATES

(following page 84)

1. Procession of Egyptians, tomb of Seti I, Giovanni Battista Belzoni, *Narrative of the Operations and Recent Discoveries within the Pyramids.*

2. Procession of "Jews," tomb of Seti I, Giovanni Battista Belzoni, *Narrative of the Operations and Recent Discoveries within the Pyramids.*

3. Figures from tomb of Seti I, C. R. Lepsius, *Denkmaeler aus Aegypten und Aethiopien.*

4. Procession of "Ethiopians" and "Persians," tomb of Seti I, Giovanni Battista Belzoni, *Narrative of the Operations and Recent Discoveries within the Pyramids.*

5. Figures from tomb of Seti I, Heinrich von Minutoli, *Reise zum Tempel des Jupiter Ammon.*

6. Eye colors, Paul Broca, *Instructions générales pour les recherches et observations anthropologiques.*

7. Hair and skin colors, Paul Broca, *Instructions générales pour les recherches et observations anthropologiques.*

FIGURES

1. Figures from tomb of Seti I, Jacques-Joseph Champollion-Figeac, *Égypte ancienne.* — 18

2. Figures from tomb of Seti I, Josiah Nott and George Gliddon, *Types of Mankind.* — 20

3. From Juan González de Mendoza, *The History of the Great and Mighty Kingdom of China*. 33

4. The animal kingdom (detail), Carl Linnaeus, *Systema naturae*. 49

5. *Homo sapiens*, Carl Linnaeus, *Systema naturae*. 55

6. Five skulls, Johann Friedrich Blumenbach, *De generis humani varietate nativa*. 63

7. Colors of races, Johann Friedrich Blumenbach, *De generis humani varietate nativa*. 64

8. Five races, Johann Friedrich Blumenbach, *De generis humani varietate nativa*. 65

9. Color top, Milton Bradley, *Elementary Color*. 91

10. Color top, Louis R. Sullivan, *Essentials of Anthropometry*. 93

11. "Slanted eyes," Philipp Franz von Siebold, *Nippon*. 104

12. The epicanthus, Friedrich August von Ammon, "Der Epicanthus und das Epiblepharon." 106

13. "Mongolian spots," Karl Toldt, "Über Hautzeichnung bei dichtbehaarten Säugetieren." 109

14. Mongolians and "Mongoloids," F. G. Crookshank, *The Mongol in Our Midst*. 119

15. "The Yellow Peril," *Harper's Weekly*. 127

16. Scale of races in America, Franklin H. Giddings, "The Social Marking System." 142

Acknowledgments

My greatest debt is to the National Science Council in Taiwan, for whose research support I am immensely grateful. I am also indebted to the superb staff at the National Taiwan University Library. This book was both begun and completed in Princeton, New Jersey, first at the Institute for Advanced Study, where I spent a blissful semester at the School of Historical Studies in the spring of 2007, and then for another blissful semester as a visiting fellow in the Department of History at Princeton University in the spring of 2009. In between I had the honor of giving a version of chapter 1 at the Shelby Cullom Davis Center for Historical Studies, also at Princeton; my thanks to Gyan Prakash for inviting me. At the IAS and at Princeton I would like to thank Heinrich von Staden, Nicola di Cosmo, Jonathan Israel, Angela Creager, Susan Naquin, Benjamin Elman, and Michael Laffan—as well as the librarians at both institutions. I also thank Brandy Alvarez, Neil Bernstein, Rae-an Chao, Bill Creager, Leif Dahlberg, Colin Day, Su-ching Huang, Virginia Jealous, the staff of the library at the Linnean Society of London, Robert Markley, Arthur Marotti, Staffan Müller-Wille, David Ogawa, David Porter, Karen Reeds, Haun Saussy, Charles Shepherdson, Matt Stanley, Jing-fen Su, Kirill Thompson, Jerry Weng, Bonnie Wheeler, Laura Jane Wey, and Wei-fan Yang.

I am very grateful to Clara Platter of Princeton University Press for taking an early interest in this project and then seeing it through to completion. Richard Isomaki was an excellent copyeditor. Richard Ho and his staff at Efoto Imaging in Taiwan were indispensible for the illustrations. And because they are exemplary colleagues as well as exemplary people, this book is dedicated to Angela N. H. Creager and Kirill Ole Thompson.

Becoming Yellow

Introduction

No Longer White

The Nineteenth-Century Invention of Yellowness

I first came to this project because I was interested in learning how East Asians became yellow in the Western imagination. Yet I quickly discovered that in nearly all the earliest accounts of the region, beginning with the narratives of Marco Polo and the missionary friars of the thirteenth century, if the skin color of the inhabitants was mentioned at all it was specifically referred to as white. Where does the idea of yellow come from? Where did it originate?

Many readers will be aware that a similar set of questions has been asked with respect to "red" Native Americans, and that the real source of that particular color term, much like East Asian yellow, remains something of a mystery. There is some evidence to suggest that the idea of the "red Indian" may have been influenced (although not fully explained) by the fact that according to European observers certain tribes anointed themselves with plant substances as a means of protection from the sun or from insects, and that this might have given their skin a reddish tinge. The stereotype of Indian war paint also comes to mind. Some tribes even referred to themselves as red as early as the seventeenth century, probably in order to distinguish themselves from both the European settlers and their African slaves.

Yet however flimsy or incomplete these accounts may be for Native Americans, in the case of East Asians there simply are no analogous explanations. No one in China or Japan applied yellowish pigment to the skin (and China and Japan will be the subject of this book; information about Korea was particularly sparse before the twentieth century), and no one in the Far East referred to himself as yellow until late in the nineteenth century, when Western racial paradigms, along with many other aspects of modern Western science, were being imported into Chinese and Japanese contexts. But yellow does have very important significations in Chinese (but not Japanese) culture: as the central color, the imperial color, and the color of the earth; the color of the originary Yellow River and the mythical Yellow Emperor, the supposed ancestor of all Han Chinese people. "Sons of the Yellow Emperor" is still in use as a means of ethnic self-identification. Could the idea of yellow people have stemmed from some form of misunderstanding or mistranslation of these symbols? Most of them were well known to early Western commentators, especially the missionaries whose aim it was to learn about local beliefs and local cultural practices for the purposes of religious conversion. Their accounts of China routinely mentioned the Yellow River and the Yellow Emperor, and it is not difficult to imagine that such symbols could have been extended to represent the cultures of the entire East Asian region, just as Chinese learning and its written language had spread beyond the confines of the Celestial Empire.

And yet in every instance in which some idea of yellow in China was analyzed or even mentioned in the pre-nineteenth-century literature, there is not a single case I am aware of in which it was connected to the color of anyone's *skin*. The idea that East Asian people were colored yellow cannot be traced back before the nineteenth century, and it does not come from any sort of eyewitness description or from Western readings of East Asian cultural symbols. We will see that it originates in a different realm, not in travel or missionary texts but in scientific discourse. For what occurred during the nineteenth century was that yellow had become a *racial* designation. East Asians did not, in other words, become yellow until they were lumped together as a yellow *race*, which beginning at the end of the eighteenth century would be called "Mongolian."

This book is therefore concerned with the history of race and racialized thinking, and it seeks to redress an imbalance in the enormous field of race studies generally, which has concentrated intently on the idea of blackness as opposed to whiteness. The few treatments of the yellow race that have hitherto appeared, such as Lynn Pan's

Sons of the Yellow Emperor or Frank Wu's *Yellow: Race in America Beyond Black and White*, have not been concerned with what we might call the prehistory of yellowness but only with its twentieth- and twenty-first-century manifestations. And texts that have provided a more historically nuanced account, such as Frank Dikötter's *The Discourse of Race in Modern China*, or his edited volume, *The Construction of Racial Identities in China and Japan*, have either sidestepped the question or given a partial and sometimes faulty summary.

The best work on the subject includes an excellent essay in German, "Wie die Chinesen gelb wurden" (How the Chinese Became Yellow), by Walter Demel, which along with an expanded version in Italian has served as the starting point for the present study. Rotem Kowner has also written suggestively on the "lighter than yellow" skin color of the Japanese, and David Mungello's introductory volume, *The Great Encounter of China and the West*, includes a short section called "How the Chinese Changed from White to Yellow." Despite such promising titles (and my own is equally guilty), each of these authors has discovered that trying to trace any straightforward development of the concept of yellowness is full of dead ends, because, as we will see in chapter 1, like most other forms of racial stereotyping, it cannot be reduced to a simple chronology and was the product of often vague and confusing notions about physical difference, heritage, and ethnological specificity.

Yet I have also followed the lead of these authors by pursuing a trajectory that emphasizes an important shift in thinking about race during the course of the eighteenth century, when new sorts of human taxonomies began to appear and new claims about the color of all human groups, including East Asians, were put forward. The received version of this taxonomical story, which we will trace in chapter 2, goes something like this. In 1684 the French physician and traveler François Bernier published a short essay in which he proposed a "new division of the Earth, according to the different species or races of man which inhabit it." One of these races, he was the first to suggest, was yellow. More influentially, the great Swedish botanist Carl Linnaeus burst onto the international scene with his *Systema naturae* of 1735, the first major work to incorporate human beings into a single taxonomical scheme in which the entire natural world was divided between the animal, vegetable, and mineral kingdoms. *Homo asiaticus*, he said, was yellow. Finally, at the end of the eighteenth century, Johann Friedrich Blumenbach, also a physician and the founder of comparative anatomy, definitively proclaimed that the people of the Far East were a yellow race, as distinct from the white "Caucasian" one, terms that have been with us ever since.

Yet there are a number of errors in this (admittedly oversimplified) narrative. In the first place Bernier did not say that East Asians were yellow; he called them *véritablement blanc*, or truly white. The only human beings he described as yellow, and not associated with an entire geographical grouping at all, were certain people from India, especially women. Immanuel Kant, also sometimes invoked as a source in this regard, agreed that Indians were the "true yellow" people. Second, we can indeed credit Linnaeus as the first to link yellow with Asia, but we need to approach this detail with considerable care, since in the first place he began by calling them *fuscus* (dark) and only changed to *luridus* (pale yellow, lurid, ghastly) in his tenth edition of 1758–59. Second, he was talking about the whole of Asia and not simply the Far East. As for Blumenbach, it is true that he unequivocally named East Asians as yellow (the Latin word he chose was *gilvus*, also revised from *fuscus*), but he simultaneously placed them into a racial category that he called the Mongolian, and it is this newly minted "Mongolianness" that has been unduly ignored in previous work on the subject.

For it was not simply the case that taxonomers settled upon yellow because it was a convenient intermediary (like red) between white and black—the two primal skin colors that had been taken for granted by the Judeo-Christian world for more than a thousand years. Rather, I would suggest that there was something dangerous, exotic, and threatening about Asia that "yellow" and "Mongolian" helped to reinforce, both of these terms becoming symbiotically linked to the cultural memory of a series of invasions from that part of the world: Attila the Hun, Genghis Khan, and Tamerlane, all of whom were now lumped together as "Mongolian" as well. While this suggestion still does not fully explain why yellow was chosen from a myriad of other color possibilities, many of which continued to be used even after

Blumenbach's influential pronouncements, yellowness and Mongolianness mutually supported each other to solidify a new racial category during the course of the nineteenth century.

Travelers to East Asia began to call the inhabitants yellow much more regularly, and the "yellow race" became an important focus in nineteenth-century anthropology, the subject of my chapter 3. Early anthropology was overwhelmingly concerned with physical difference in addition to language or cultural practice, and skin color was one such preoccupation. Blumenbach and the comparative anatomists were obsessed with the measurement of human skulls, producing a theory of "national faces" that led to a hierarchical arrangement of the symmetrical "Caucasian" shape as opposed to more lopsided forms manifested by the other racial varieties. Blumenbach and his followers placed "Mongolian" skulls, along with "Ethiopian" ones, at the furthest extreme from the Caucasian ideal, with "American" and "Malay" heads in between.

But as anthropology came into its own in the middle of the nineteenth century, the process of physical measurement became much more complex and extended to minute quantifications of the entire body. A key figure here was Paul Broca, who by the time of his death in 1880 had invented more than two dozen specialized instruments for the purposes of human measurement. Less well known is his highly influential foray into the assessment of skin color, which, as we will see, he attempted to standardize by developing a chart with colored rectangles designed to be held up to the skin in order to find the closest match. Others tried to improve upon this rather cumbersome and subjective procedure by experimenting with different color ranges and introducing different media, such as glass tablets or oil paints, and by the end of the nineteenth century one popular alternative was a small wooden top upon which were placed a number of colored paper disks that blended together when the top was spun. The subjects to be measured would rest an arm upon a table next to the spinning top while the researcher adjusted the disks until they matched the color of the skin.

Such methods may seem quaint or entertaining today, but anthropologists took them very seriously and used them with great frequency in many parts of the world. What especially interests me, however, is the way in which these tools functioned as means to invest preexisting racial stereotypes with new and supposedly scientifically validated literalness. Colors on color charts were never chosen and organized arbitrarily, and the color top employed white, black, red, and yellow disks despite the fact that many other combinations could have been used to replicate the rather limited tonal range that comprises human skin. This was not because, as the top's early developers claimed, these were the pigments actually present in the skin. Rather, white, black, red, and yellow were the colors presumed for the "four races of man" from the outset. When researchers began to quantify "Mongolian" skin color, it turned out to be some sort of in-between shade between white and black, and when the dice were loaded carefully enough, as in a color top, East Asian skin could turn out to be yellow after all.

In chapter 4 we will perceive a similar development in nineteenth-century medicine, which instead of color focused on the quantification of "Mongolian" bodies by associating them with certain conditions thought to be endemic in, or in some way linked to, the race as a whole, a list that includes the "Mongolian eye," the "Mongolian spot," and "Mongolism" (now known as Down syndrome). I will argue that each of these conditions became a way of distancing the Mongolian race from a white Western norm, since they were taken to be either characteristic of irregular East Asian bodies, as in the case of the Mongolian spot, which did not seem to occur among white people at all, or a feature that appeared among whites only in their youth or if they were afflicted by disease, as in the case of the Mongolian eye or Mongolism. Researchers also linked these "Mongolian" conditions to contemporary evolutionary theories about the way in which the white race had passed through the developmental stages still occupied by the lower ones. Thus the Mongolian spot, which was first noticed on Japanese babies, was seen as a pigmentary trace of an earlier stage of human evolution, perhaps even the trace of a monkey's tail; white children might have something very like a Mongolian eye before they simply outgrew it; and people with Mongolism, especially children, resembled racial Mongolians because it was a visible throwback to a previous evolutionary form.

Much as in the case of early anthropology, medical explanations for "Mongolian" pathology had an uncanny way of reinforcing the stereotypes with which researchers began. Physicians, too, regularly described East Asians as having yellow bodies, but "Mongolian" conditions could be linked to physiological degeneration and play into even older clichés about the static, infantile, and imitative Far East. White people might be afflicted with "Mongolian" traits temporarily or because of ill-health or a birth defect, but real yellow people remained stagnant and frozen in a permanent state of childishness, subhumanity, or underdevelopment.

By the end of the nineteenth century modern science had fully validated the yellow East Asian. But this yellowness had never ceased to be a potentially dangerous and threatening racial category as well, becoming particularly acute after larger numbers of East Asians had actually begun to immigrate to the West starting in the middle of the nineteenth century. The Far East now came to be seen as a "yellow peril," a term coined in 1895 and generally credited to Kaiser Wilhelm II of Germany, specifically in response to Japan's defeat of China, its far larger and more populous neighbor, at the conclusion of the Sino-Japanese War, also known as "The Yellow War." Even worse, Japan had begun to form a colonial empire of its own, and when ten years later it had defeated Russia, too, it seemed to mark the end of the West's control of the civilized world. This period will occupy us in chapter 5.

The yellow peril was a remarkably free-floating concept that could be directed at China or Japan or any other "yellow" nation, as well as to many kinds of perceived peril such as overpopulation, "paganism," economic competition, and societal or political degradation. But we will also see that the West had begun to export its purportedly self-evident definitions of yellowness and Mongolianness into East Asian contexts, and that this dispersal was hardly simple and straightforward. In China, where yellow was such an ancient and culturally significant color, the West's notion of a yellow race was a happy coincidence and could be proudly inverted as a term of self-identification rather than just a racial slur, and not simply a cultural symbol but the actual color of Chinese nonwhite, non-Western skin. "Mongolian," however, linked to the non-Chinese "barbarians" who had historically been the bane of China as well as the West, was summarily rejected. Japanese commentators, on the other hand, disavowed both yellow and Mongolian, which were said to be descriptors of other Asians only, especially the Chinese. Many Japanese preferred to be considered closer to the powerful white race than the lowly yellow one, and indeed many in the West agreed. In both China and Japan, however, Western racial paradigms had become so pervasive that even those for whom "yellow" was a term of opprobrium begrudgingly admitted that their skin color was something other than white.

I will bring this story to an end in the early twentieth century, not because it ceases to be interesting or important, but simply because after the 1920s and 1930s the idea of a yellow race—and of race in general—would be much better suited to a separate study. By those decades a yellow and a racially Mongolian Far East had crossed boundaries of language, discourse, location, level of education, and social rank (as well as boundaries of gender, not pursued in this book). I also do not attempt to trace similar developments in the representation of yellow people in the vast realm of literary, visual, and other arts (fiction and satire, political cartoons, book illustrations, chinoiserie objects, Hollywood film, vaudeville and stage plays, music). As was broadly the case with travel or scientific descriptions, artistic depictions of the people of the Far East were not yellow until (at the very earliest) the beginning of the nineteenth century.

Concluding in the early twentieth century, moreover, will help to emphasize what has hitherto escaped the full attention of scholars in the history of race and in East-West cultural studies generally, and what should be much more carefully examined in conjunction with the twentieth- and twenty-first-century forms of prejudice that are still being felt so acutely today. First, the idea of a yellow-colored people, centered in Asia, was new in 1800. Second, at much the same time, this notion of a racial group began to migrate away from Asia as a whole, itself a profoundly slippery and mythic Western geographical category, and toward what we now refer to as East Asia. And third, the catalyst for both these developments was the invention of a Mongolian (and later, Mongoloid) race.

THE YELLOW FACE OF SATAN

In order to emphasize the way in which yellow was new at the turn of the nineteenth century, and that it pervaded fields of inquiry that might seem far removed from questions of race, I would like to begin with two examples. The first is a well-known passage from the last canto of Dante's *Inferno*, in which, in the ninth and last circle of hell the poet sees a terrifying vision of Satan, who is described as having three faces:

> Oh what a great marvel appeared to me
> when I saw three faces on his head!
> The one in front, this one was vermilion;
>
> and there were two others, joined with this one
> above the middle of each shoulder
> and joined at the crest:
>
> the right one seemed between white and yellow;
> the left was such to look upon as those
> who come from where the Nile descends.
> *Inferno* 34.37–45

Accustomed as we are to visualizing the world according to racial categories, it is not hard to imagine modern readers wondering whether these colored faces were supposed to represent different forms of human ethnicity. The left face, not actually named but situated geographically as "where the Nile descends," is usually taken as black (although of course black is an equally imaginary skin tone and not all Africans are dark). But what of the other two faces? Why is the central face red or vermilion, and the one on the right "between white and yellow"?

Some readers might wish to appeal to the precedent of early European world maps, which did indeed take a tripartite form. Known as T-O maps, they were shaped like a large T inside a circle, with the Asian continent situated on top and Europe and Africa, to the left and right respectively, placed beneath. And yet these maps never organized the world according to skin color, although Europeans certainly did think of themselves as white in contrast to most Africans, and although it was also something of a medieval stereotype (as in Isidore of Seville's *Etymologies*) that the inhabitants of the Indies—which meant all the lands to the east of Europe—were "tinged with color" owing to the burning heat of the region.[1]

To put the matter as succinctly as possible, the notion that Satan's three faces should be read as a reference to race has nothing whatever to do with Dante. I do not know of a single example of this sort of pigeonholing until the seventeenth century, and even then it was very exceptional. It was no accident, however, that racialized readings began to appear when obsessions with skin color classifications were increasingly becoming the norm. The first such reader seems to be Baldassare Lombardi, whose landmark 1791 edition of the poem created something of a stir when it attempted to solve the crux of Satan's three faces in a completely new way. His commentary began by noting that there had been considerable dispute about what the colors signified, but that it has been generally agreed upon that since Satan was a monstrous inverse or an ironic perversion of the triune God, it was appropriate for him to have three faces showing wrath (vermilion), envy or avarice (yellow-white), and sloth (black). Indeed in the fourth book of *Paradise Lost* Milton would echo this passage when he characterized Satan's face as "thrice chang[ing] with pale ire, envy, and despair."

But "according to me," Lombardi announced, "it might be better to understand these three faces and their colors as corresponding to the three parts of the world as the poet knew it in his day, that is, Europe, Asia, and Africa, in order to indicate that Lucifer is master of every part of the globe." Dante almost certainly conceived of the world as consisting of three continents, and he would have agreed that Satan's dominion extended to all human beings, afflicted as they are by original sin. Yet how do Europe, Asia, and Africa correspond to these particular colors? Lombardi's explanation, in fact, was breathtaking in all its banality, but it was also revealing of a new conception of the world as constituted not only by people of different nations, regions, and cultures, but indeed by people who were to be distinguished by the different colors of their skin.

Europeans were vermilion, he explained, because this is what "the majority of them have in their faces."[2] At first glance this might seem a rather puzzling claim, despite well-worn literary clichés about rosy-cheeked heroes or the beautiful white-and-red faces of Petrarchan ladies. But Lombardi might also be alluding to new taxonomical schemes such as Blumenbach's, which when it was definitively revised in 1795 would remark that the primary color of human beings was white, and that this whiteness could be properly identified by a "redness of the cheeks rarely found in the other [varieties]." In his *History of the Earth* of 1774, Oliver Goldsmith had similarly written that the complexion of Europeans was the most beautiful because "every expression of joy or sorrow flows to the cheek." "The African black and the Asiatic olive complexions admit of their alterations also," he added, "but these are neither so distinct nor so visible as with us; and in some countries the colour of the visage is never found to change."[3]

In addition to being white, in other words, Europeans were also, physiologically speaking, "the blushing race," and the popularity of this term grew when in 1845 an Old Testament scholar even claimed that the Hebrew word for Adam signified "the red." Such claims were common throughout the nineteenth century; as late as 1924, in a reprint of Edward Tylor's standard introduction to anthropology first published in 1881, there appeared an assertion that the difference between the light and the dark races was "well observed in their blushing."[4]

So far so good: Lombardi was able to explain the vermilion face, the face in front, as an allegorical representation of Europeans. And yet there was still the question of that other "between white and yellow" visage, which presumably should represent the third continent, Asia. Worse still, there was no biblical or other ancient authority to fall back on, and travelers' reports had characterized Asians with a staggering variety of different color terms. Goldsmith, for instance, called them olive. Lombardi could only put forward the supposedly self-evident fact—which once again was entirely new—that like blushing Westerners and black Africans this was simply the color that Asians really were. "The yellow-colored face," he continued, represents "the people of Asia on account of the great number there who are of such a color."[5]

Although blind to the outrageous historical anachronism of this sort of explanation, later interpreters continued to repeat it, even if cautiously. An excellent example is Dorothy Sayers's once standard English translation published by Penguin in 1949. "The three faces, red, yellow, and black," she noted, "are thought to suggest Satan's dominion over the three races of the world: the red, the European (the race of Japhet); the yellow, the Asiatic (the race of Shem); the black, the African (the race of Ham)." In addition to fixing these colors as definitively red, yellow, and black, confidently assumed to be "the three races of the world" in Dante's time, Sayers also suggested a different but commonly argued medieval tradition stemming from the tenth chapter of Genesis, that the races of the earth originated in the sons of Noah (although this is not actually mentioned in Dante's passage). Dark-skinned people, for example, were regularly referred to as being marked with "the curse of Ham." Sayers was not necessarily endorsing these readings, we should note, and she helpfully added an alternative one in which the faces were also "undoubtedly a blasphemous anti-type of the Blessed Trinity."[6]

Explicitly racial readings of the passage have since fallen out of favor. But they have certainly not disappeared either, as in Mark Musa's 1996 commentary, where the racial interpretation is not sufficiently explained as unhistorical, and, much worse, as in Elio Zappulla's 1998 translation published by Pantheon, where the footnotes state without comment that "the three colors of the faces may symbolize the races of humanity."[7] But I would argue that the history of this passage reveals the particularly difficult problem of fixing *Asians* according to any sort of rigid color scheme. It might have seemed easy enough to find a certain kind of precedent for vermilion Europeans, and Africans had been thought of as dark or black since at least the beginning of the Christian era. But what about a yellow

Asia, not just East Asia but Asia as a whole? Early nineteenth-century readers seized upon it as a means of retrospectively fixing Asian color once and for all, as if the truth of its yellowness had been just as routine at the end of the thirteenth century as it was beginning to be some six hundred years later. Yet it was only at the beginning of the nineteenth century that yellow had been selected from the myriad of other equally (in)appropriate candidates, and it was only at this time that appeals to a yellow Asia could seem every bit as obvious as a black Africa or a blushing Europe. In other words, the most revealing aspect of Lombardi's reading, as well as of those that have continued to mention it, is that it was based on nothing at all.

YELLOW ANCIENT EGYPTIANS

A similar fate is embodied by our second example, which was even more attractive to the early nineteenth century since it was explicitly visual. It concerns the tomb of the Egyptian pharaoh Seti I, discovered in 1817.[8] Located in the Valley of the Kings in the city of Thebes, the tomb dates from the 19th Dynasty of the Egyptian New Kingdom and was built in the late thirteenth century B.C. Owing to its enormous size, depth, and plethora of wall decoration, including magnificently colored paintings in raised relief, it was easily the most spectacular and widely known Egyptian site until the discovery of Tutankhamun's tomb in the 1920s. But for early nineteenth-century admirers one set of paintings immediately stood out from the rest, since they seemed to show not only that ancient Egypt had been just as preoccupied with racial difference as was the modern West, but also that the races had actually been depicted according to a very similar system of skin colors.

These paintings are situated in a large pillared chamber depicting a procession, among which appear small groups of men carefully differentiated in terms of their costume, body ornament, headgear, and hairstyle (color plate 3). There are four such groups, including a party of Egyptians themselves, who might be represented as returning prisoners of war. But what really attracted nineteenth-century viewers is that these men were also differentiated in terms of their color. As was customary in Egyptian self-representations the men were depicted using a red pigment (Egyptian women, incidentally, were generally shown as yellow), but the foreigners were endowed with both lighter and darker tints, and in his narrative of the discovery published in 1820, Giovanni Battista Belzoni identified them as "evidently Jews, Ethiopians, and Persians." A gargantuan folio of plates also provided lithographs copied after watercolors that were executed on site (color plates 1, 2, and 4).

The vagaries of tourism, plundering, humidity, and dirt have since deteriorated these paintings to such a degree that it is nearly impossible to verify many of the details that Belzoni and his contemporaries claimed to see, including, most notably, the colors. But the paintings became famous at once and were repeatedly touted, much like the passage from Dante, as representations of human ethnicity. And indeed they are. But in the nineteenth century they were treated almost as race samples in a contemporary anthropological textbook, despite the fact that it was far from clear which groups were being depicted and, even more importantly, how that information would have been appreciated by an ancient Egyptian audience. Moreover, since the hieroglyphic tags were imperfectly understood, Belzoni had to rely on other details such as clothing and ornamentation, and as in the case of Satan's three faces, readers began to claim familiar and supposedly self-evident racial traits.

"The Jews are clearly distinguished by their physiognomy and complexion," Belzoni tersely noted, "the Ethiopians by their colour and ornaments, and the Persians by their well-known dress." An anonymous remark included at the end of Belzoni's narrative agreed, but this author placed even more emphasis on skin color: "red men with white kirtles" (Egyptians), "white men with thick black beards" (Jews), "negroes with hair of different colours" (Ethiopians), and "white men with smaller beards" (Persians). More revealingly, it was also claimed that the figures of the foreigners "exhibit the most remarkable feature of the whole embellishments of the catacomb," even though there were so many other magnificent paintings spread across numerous rooms, entryways, corridors, and stairwells, including the enormous burial chamber and its dazzling alabaster sarcophagus, which attracted huge crowds when it ended up on display in London along with other artifacts and a detailed model of the entire tomb.⁹

As in the passage from Dante, however, an explicitly racial reading did not accord very well with nineteenth-century assumptions. The black faces, as always, were immediately identified with Africa, but the Egyptians were red and there were now two white races, Jews and Persians. A solution would soon be offered when the site was visited by Jean-François Champollion, famous for being the first to make significant progress in the deciphering of hieroglyphics. Armed with the ability to read the texts, he suggested identifications that were slightly different, but he also managed to convince himself that ancient Egyptian racial categorizations were essentially indistinguishable from those of the nineteenth century. "We have here before us," he asserted, "the image of diverse races of men known by the Egyptians"; "here are figured the inhabitants of the four parts of the world according to the ancient Egyptian system."¹⁰ In other words, it was quickly supposed not simply that these were people with whom Seti had had contact or whom he had vanquished, but indeed that they formed a kind of symbolic tableau of the races of the world, much as Satan's three faces were said to be plotted on a race-based map.

Once again this notion was not entirely misguided. The paintings appear in the context of a depiction of a funerary narrative known as *The Book of Gates*, which frequently included representations of people from other tribes or nations and from each of the four directions to emphasize that they, too, were sheltered in the realm of the dead.¹¹ A scholar versed in hieroglyphics would naturally be better equipped to place the figures into such a framework, and similar groups were also featured in the nearby tomb of Rameses III, which Champollion also mentioned, and which had been fully opened to tourists and adventurers for more than fifty years before Belzoni had arrived.

But the figures in Seti's tomb were in a far better state of preservation and, for Champollion, *plus véritable* in their racial differentiation. Indeed, as one reads through his description one can almost see him straining to make them conform as closely as possible to the desired goal of clear and distinct white, black, and yellow peoples. The Egyptians were the center of this universe, identified with the hieroglyphic tag "Rôt-en-ne-Rôme, the race of men, men par excellence." They embodied "a dark red color, [are] well proportioned, with a soft physiognomy, slightly aquiline nose, long braided hair, and dressed in white." The second group, whom Belzoni had identified as Jews, were now called "Asiatics" and labeled "Namou." Even more strangely, however, they had also become yellow—or, rather, a similar kind of "white and yellow" that had characterized the "Asian" face of Satan. The "Namou," he wrote, "present a very different aspect: flesh-colored skin tending to yellow or a swarthy hue [*peau couleur de chair tirant sur le jaune, ou teint basané*], a strongly aquiline nose, a black and abundant pointed beard, and short vestments of various colors." The next group, "about which there can be no uncertainty," were *nègres* and were designated by the name "Nahasi." Last were Belzoni's Persians, whom Champollion labeled "Tamhou," and they had undergone the most startling transformation of all. For these faces were now said to be "flesh-colored or white-skinned with the most delicate nuance [*couleur de chair, ou peau blanche de la nuance la plus délicate*], the nose straight or slightly arched, blue eyes and blonde or red beards, of tall and willowy stature, dressed in cowhide that still retains the hair, veritable savages tattooed on various parts of their bodies." For Champollion, that is, this other white race was fantasized not merely as a people from beyond the northern borders of the Egyptian state, but indeed as a representation of straight-nosed, blonde, blue-eyed, and elegant Europeans. While they might be shown as tattooed barbarians ("I am ashamed to say it," he admitted, "since our race is the last and the most savage of the series"), these men were just as clearly "a race apart": "our beautiful ancient ancestors."[12]

After his death in 1832, Champollion's identifications were then integrated into an introductory volume on Egypt composed by his elder brother and published as part of the popular series *L'univers*. Here the problem of red Egyptians was directly addressed, just as the whole question of race was also incorporated into a larger and hotly debated discussion about precisely what color the Egyptians were. Were they really black Africans, as the ancient Greco-Roman world had considered them to be? This question was of considerable importance if they were to assume their rightful place as an originary

Western civilization. Blumenbach had argued, after examining the evidence of mummies that had been brought to London at the end of the eighteenth century, that they were indeed black, or, according to his particular hierarchy of the races, midway between Caucasians and Ethiopians. A somewhat different theory, dating from at least the middle of the seventeenth century, had claimed that they were related to Asiatics as far away as China, which might even have originated as an Egyptian colony. This was thought to explain the marked similarities between the two cultures, especially their mysterious pictorial languages. Winckelmann's *History of Ancient Art*, first published in 1764, echoed these ideas when he noted that "statues, obelisks, and engraved gems show that the form peculiar to [the Egyptians] somewhat resembled that of the Chinese." But Blumenbach countered this view as well; the Egyptians, he wrote, "differ from none more than from . . . the Chinese."[13]

But as many of his contemporaries were to do, the elder Champollion argued that the Egyptians were Moorish and not black at all, and that they differed from Europeans only in that their skin had been "browned" by the climate—a fact that could supposedly be proved by the "well proportioned," "soft," and "aquiline" figures shown in the tomb. His book also included a new engraving showing the six "Peuples connus des Égyptiens": four figures from Seti's tomb and two additional representations of a "Persian" and a "Greek" taken from other sites (figure 1). Skin colors were not delineated here, with the exception of a hatched black figure, perhaps in order to emphasize that the Egyptians were just as white as the others. And regarding the Egyptian figure, specifically, "it is impossible to find . . . any of the traits that characterize the *race nègre*. The facial angle is beautiful, the features are regular, the lips pronounced but well joined, and the rest of the body having a comportment that one recognizes in individuals of the white race."[14]

Appeals to such quantitative measurement as facial angle, first proposed by the Dutch anatomist Petrus Camper in the late eighteenth century and developed by J.-J. Virey and J. B. Bory de Saint-Vincent in the nineteenth, were quickly becoming a standard feature of new taxonomies of racial difference. Samuel George Morton went

Figure 1: "Peoples Known to the Egyptians," from Jacques-Joseph Champollion-Figeac, *Égypte ancienne* (1839). With the exception of one dark figure these representations were not differentiated in terms of their skin color. Princeton University Library.

even further when he claimed that human intelligence could actually be measured by skull capacity, and in his volume on the Egyptian evidence, published as *Crania Aegyptiaca* in 1844, he agreed that they were not African. By the time of Henri Brugsch's *Histoire d'Égypte* of 1859, the "grave and important" question of the origin of the Egyptian people had been (for Brugsch, at least) sufficiently solved: they were now firmly part of the Caucasian race. Henceforth red Egyptians (and yellow Egyptian women) would rarely be interpreted in any literal way, since this was not really their correct skin tone.[15]

But the problem of two apparently white races remained, exacerbated by the fact that their skin color varied considerably in nineteenth-century reproductions. In Belzoni's lithographs the Jews did seem to differ from the Persians and could indeed be called yellowish, but in a competing watercolor produced by Heinrich von Minutoli in 1820, also executed on site and first published in 1827, the hue of the two light-skinned figures was indistinguishable (color plate 5). The same could be said for the lithographs published by Ippolito Rosellini (who had traveled with Champollion) in 1832. Yet when they appeared in C. R. Lepsius's *Denkmaeler aus Aegypten und Aethiopien*, the result of new expeditions in the early 1840s, the Asians were shown (and indeed described) as yellow-brown and far darker than the Europeans, who were now privileged as the only true white race in the world (see color plate 3).[16] While the Egyptians themselves could not really be red, black Africans, yellow Asians, and white Europeans were thought to be represented with complete accuracy, so much so that by the end of the nineteenth century the paintings in the tomb took on an aura of almost photographic verisimilitude. The Asian faces were now routinely said to *be* yellow, and one particularly imaginative reader, in an essay published in the inaugural volume of the *Annales du Musée Guimet* in 1880, even claimed that those depicted as European had been given *une teinte rosée*—thus verifying that even in the thirteenth century B.C. they were already being recognized as "the blushing race."[17]

But the real crux for nineteenth-century readers, I would argue once again, was how to colorize the intermediate races that were neither white nor black. Red Egyptians could be easily ignored or explained away, but two apparently white peoples needed to be carefully differentiated. Josiah Nott and George Gliddon's best-selling *Types of Mankind* of 1854 solved this problem by reproducing a plate of the "four species" in which the nonwhite faces were deliberately shaded as three distinct skin tones. Lest we miss the point, the caption clearly identified them as red, yellow, black, and white (figure 2). The paintings, in other words, were now distinguished *solely* in terms of color, but Nott and Gliddon also revealingly pointed out that it was actually the yellow faces that had previously stood in their way. Earlier reproductions, they noted, had not properly distinguished the white from the yellow: "we were always at a loss to account for the presence of *two* white races in Rosellini's copy of this tableau. It turns out that an error of *coloring* on the part of the Tuscan artists was the unique cause of such perplexities; because they have tinted this figure *light flesh-color*, instead of a tawny yellow."[18]

I should point out that the original artists may indeed have chosen to represent the "Namou" with a yellowish pigment, although judging by the figures in the tomb of Rameses III, it is difficult to distinguish

ON TYPES OF MANKIND.

Fig. 1.
The ancient Egyptian division of mankind into four species—fifteenth century B. C.

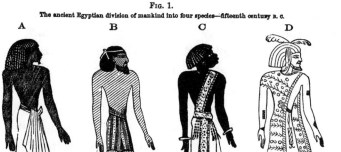

Red.　　　　Yellow.　　　　Black.　　　　White.

The above figures, which may be seen, in plates on a folio scale, in the great works of Belzoni, Champollion, Rosellini, Lepsius, and others, are copied, with corrections, from the smaller work of Champollion-Figeac.[27] They display the *Rot*, the *Namu*, the *Nahsu*, and the *Tamhu*, as the hieroglyphical inscription terms them; and although the effigies we present are small, they portray a specimen of each type with sufficient accuracy to show that *four* races were very *distinct* 3300 years ago. We have here, positively, a scientific *quadripartite* division of mankind into *Red*, *Yellow*, *Black*, and *White*, antedating Moses; whereas, in the Xth chapter of *Genesis*, the symbolical division of "SHEM, HAM, and JAPHET," is only *tripartite*—the Black being entirely omitted, as proved in PART II. of this volume.

The appellative "*Rot*" applies exclusively to one race, viz., the *Egyptian;* but the other designations may be somewhat generic, each covering certain groups of races, as do our terms Caucasian, Mongol, &c.; also including a considerable variety of types bearing general resemblance to one another in each group, through shades of color, features, and other peculiarities, to be discussed hereafter.[28]

EXPLANATION OF FIG. 1.

A.—This figure, together with his three fac-simile associates, extant on the original painted relievo, is, then, typical of the *Egyptians;* who are called in the hieroglyphics "*Rot,*" or Race; meaning the Human race, *par excellence*. Like all other Eastern nations of antiquity—like the Jews, Hindoos, Chinese, and others—the Egyptians regarded themselves alone as the chosen people of God, and contemptuously looked down upon other races, reputing such to be Gentiles or outside-barbarians. The above representation of the Egyptian type is interesting, inasmuch as it is the work of an *Egyptian* artist, and must therefore be regarded as the Egyptian ideal representation of their own type. Our con-

Figure 2: "The ancient Egyptian division of mankind into four species," from Josiah Nott and George Gliddon, *Types of Mankind* (1854). The four figures (or rather "species") from Seti's tomb are now assigned clear and distinct color labels. Princeton University Library.

their color from that other white race, the "Tamhou." In any case, this has nothing to do with the claim three thousand years later that all the peoples of Asia *were* in fact yellow. Presumably, no further proof was required. By the 1880s there would be calls for more evidence to decide the facts once and for all, utilizing the new scientific tool of photography. An 1887 address to the Anthropological Institute of Great Britain concluded by urging its members to obtain "correct photographs of the portraiture of different races still remaining on the walls of the monuments before these most valuable records shall be lost to us for ever." One might well wonder why this particular aspect of Egyptian art was seen as so important, and in 1912 a major "Fremdvölker-Expedition" was actually carried out, the result being a visit to seventeen different tombs and a collection of nearly 800 photographs, most of which were published in Walter Wreszinski's *Atlas zur altägyptischen Kulturgeschichte*. But since everyone was predisposed to see the Asians as yellow, yellow they remained, as in Paul Topinard's influential *Éléments d'anthropologie générale* of 1885, where the images in Seti's tomb were said to show yellow Asians with aquiline noses and referred to Nott and Gliddon's *Types of Mankind* for support.[19]

A total collapse of historical specificity was now complete. Nott and Gliddon had remarked that "the ancient Egyptians had attempted a systematic anthropology at least 3500 years ago," and that "their ethnographers were puzzled with the same diversity of types . . . that . . . we encounter in the same localities now." Alexander Winchell's *Preadamites* of 1880 affirmed "the very high antiquity of the racial distinctions existing in modern times," and in 1909 A. C. Haddon's *Races of Man* claimed that ancient Egyptian artists "distinguished between four races" just as "we ourselves speak loosely of white men, yellow men, black men or 'niggers,' red men, and so forth." As late as 1990, Spencer Rogers's *Colors of Mankind* reproduced a fuzzy version of Nott and Gliddon's plate and noted, apparently with complete confidence, that "the artists painted figures representing the Egyptians red, the Semites yellow, the Negroes black, and the Mediterranean peoples white."[20]

As in the Dante passage the reception of these paintings glaringly demonstrates the way in which early nineteenth-century (and later) readers regularly projected their own racial preoccupations onto earlier periods as well as onto other cultures, and that one of the assumptions of this mode of thinking was that Asians were yellow. But what we have not yet understood is how that yellowness gradually became a feature associated with *East* Asians and not Asia as a whole. In the chapters that follow we will be questioning not only the history of that color term but how it became a designation for the "Mongolian" Far East. Let us begin, then, by taking stock of exactly how East Asians were described before they became yellow at the end of the eighteenth century.

Chapter 1

Before They Were Yellow

East Asians in Early Travel and Missionary Reports

When premodern European authors attempted to describe the residents of other lands there was often little agreement about precisely what color they were, partly because before the end of the eighteenth century there was no systematic desire to classify people according to what we now call race. Western thinking had long differentiated between the peoples of the known world in a variety of ways, including often vague notions about skin tone. But markers such as religion, language, clothing, and social customs were seen as far more important and meaningful than the relative lightness or darkness of the inhabitants, which, in any case, was usually attributed to some combination of climate, gender, and social rank. Human "blackness" was a conceptual marker of difference that from an early date could be associated with dirt or evil (Satan was perceived as the only truly black individual), but in a broader sense it was also an adjective that was constantly utilized to suggest sin, idolatry, and cultures that lay outside the Christian community. Anyone beyond (or on the fringes of) Europe could be coded as "dark" or "black." Yet this, too, was not a racial distinction in the modern sense of the term.[1] Even earlier, in the Greco-Roman world, skin color seemed to have held much less significance, although the lands to the east, known collectively as India, were always associated with marvels, fabulous wealth, and various forms of both human and nonhuman monsters.[2]

It is against this background that we must understand a certain level of surprise when the earliest medieval travel narrators, such as Marco Polo at the end of the twelfth century, referred both to the leader of Cathay and the people of Japan as "white" (*bianca*). This description was extended to all Chinese when a version of Polo's text (and there were many) was edited for G. B. Ramusio's pioneering travel compendium in 1559.[3] Other travelers to China such as Friar Odoric in the 1330s remarked that the people of the region were good looking (*di corpo belli*), although here the southern Chinese were described as pallid (*pallidi*) rather than white—and this, too, would become an important nuance in later accounts.[4]

Beginning at the end of the fifteenth century, when (at first Iberian) travelers sailed around the tip of Africa and into the Indian Ocean, they were pleased and gratified to find that the people of Asia were not uniformly dark. It was another medieval stereotype (as in Isidore of Seville's *Etymologies*) that the inhabitants of the Indies were generally "tinged with color" (*tincti coloris*) owing to the burning heat of the region.[5] This notion had become linked to another old legend that somewhere on the other side of the Arab world there existed a "lost" Christian (and perhaps "white") community led by the figure known as Prester John. He had been the subject of a fictional letter to the pope in 1164, asking for assistance in resisting the Arab enemy. Early modern Asian exploration as a whole might even be viewed as an attempt to fulfill a long-term obsession with finding Prester John, whose location constantly shifted as each new area became better known to Western travelers.[6]

In 1511 the Portuguese were able to establish a permanent outpost for East Asian trade at Malacca, which for some time had already been a site of thriving international commerce. Persistent rumors of "white" people in the Far East had suddenly become a reality, as both Chinese and Japanese (as well as Arabs and other East Asians) were a common sight. The "whiteness" of these people was constantly highlighted, not only in contrast to the Indians but as a term that described their presumed level of civilization. A revealing case in point is one of the earliest accounts of the arrival of the Europeans by Girolamo Sernigi, a Florentine merchant who had been in the employ of the Portuguese during the first voyage of Vasco de Gama in 1497–99. Before sailing into the Indian Ocean the Portuguese had arrived at Calicut on the southwest coast of India, where they were told about a visit some eighty years previously of "certain vessels of white [*bianchi*] Christians, who wore their hair long like Germans, and had no beards except around the mouth, such as are worn at Constantinople by cavaliers and courtiers." Sernigi added that if these sailors had really been Germans the Portuguese would have known about them, so he wondered whether they were Russians instead.[7]

We have no idea what was really said to de Gama and his party, and one must assume that the comparison to Germans was merely part of the way in which the rumor was received or retold from a Western perspective. Second, one suspects that the identification of these men as Christian was also a European assumption, since for centuries they had already fashioned an equation between human whiteness and Christianity. Moreover, the whole story was related at second- or thirdhand at best, and it was not published until 1507, long after Sernigi had returned to Lisbon. It has been plausibly argued, however, that these "white Christians" were in fact Chinese, members of an enormous seagoing operation headed by the eunuch Zheng He (or Chung Ho), who before he died in 1435 had established the Chinese as the leading international trading presence in the entire Indian Ocean region, a position they held for many decades before circumstances led to their almost complete (official) withdrawal by the time the Portuguese arrived. It has even been suggested that initially the Indians of Calicut may have welcomed the Portuguese precisely because they thought they were Chinese.[8]

Whatever the case, Sernigi's anecdote was later canonized in Luis de Camões's sixteenth-century national epic, *The Lusiads*, where it was integrated into a glorified narrative of Portuguese trade and the triumph of Christian civilization. The entire tone of the story has been altered, too, as Camões transformed Sernigi's puzzlement and suspicion into a good omen for the joyous prospect of future profit in an already well-formed system of multinational trade. When the Portuguese met dark-skinned Arabic-speaking men, Camões described them as "Ethiopians" who "had communicated with better people," and when the Portuguese also learned that "from the East" there were other ships as large as their own and manned by white men—"people much like us, of the color of day" (*Gente, assi como nós, da côr do dia*)—the implication seemed to be that these were Arab traders rather than European competitors.⁹

While the Indian informants at Calicut were unable or uninterested in distinguishing one light-colored skin from another, Sernigi and Camões both presumptuously supposed that the Europeans were the only truly "white" people in the world, which is to say the only civilized Christian nations. Darker-skinned peoples might have learned to speak Arabic and have mastered the art of overseas navigation, but it was also assumed that if there really were white men in the East they must have been Europeans like themselves, and if they were northern Africans they might also have relatively fair skin but they were certainly not "white." The European reception of this new body of information, in other words, was immediately read according to presuppositions about human culture as it was related to skin color, and while we are unable to say with any certainty whether the earlier white visitors were indeed Chinese (and the evidence seems to be based mainly on the kind of weapons they are described as carrying), this is a perfect example of Westerners' misunderstanding of their own presumed superiority.

In the version of Sernigi's letter published by Ramusio the visitors' whiteness and their imagined nationality went unmentioned, and yet the structure of European self-interest remained the same, as the desire to "find" white Christians in the East, or to Christianize them if they were of another religion, were central features of Western encounters with native peoples. King Manuel I of Portugal's oft-quoted instructions to Diogo Lopes de Sequeira in 1508, three years before the Portuguese had established their position at Malacca, was intensely concerned about the religion of (in this case) China. "Are they Christian or heathen; do any Moors live among them or others who are not of their faith; what do they believe in if they are not Christians?"¹⁰

WHITE EAST ASIANS

In the beginning of the European "age of exploration," in short, East Asian peoples were almost uniformly described as white, never as yellow. The surviving literature is full of references to the whiteness of both Chinese and Japanese natives, as merchants and (later) missionaries were able to penetrate into the mysterious lands of the marvelous East. A few examples should suffice. One of the very first accounts was by Tomé Pires, a Portuguese apothecary who stayed in Malacca between 1512 and 1515 and who compiled a long report known as the *Suma oriental*, addressed to King Manuel I. Like all such information his notes were a closely guarded secret, since the Portuguese hoped to retain a monopoly on the enormously profitable trade to the region. A part of the text was published (also by Ramusio) when that monopoly was no longer secure, including information that the Chinese were "white like us [*bianchi, si come siamo noi*], the greater part of them dressing in cotton cloth and silk." The full text compared them to Germans (something of a cliché in this period), and the women, who were also described as being "of our whiteness" (*da nosa alvura*), looked like Spanish ladies.[11]

A contemporaneous report by Duarte Barbosa, a Portuguese official in India for many years, was also published in abridged form by Ramusio. Here, too, the people of China were fair-skinned: "great merchants, white men and well-made [*huomini bianchi, grandi & ben disposti*]; their women are very beautiful but both the men and women have small eyes, and the men's beards contain only three or four hairs and no more." Once again the appearance of the people was measured according to thoroughly European standards; they were praised for wearing shoes and stockings and were compared to Germans, this time in terms of their language, as if it were an equally outlandish tongue to Iberian ears.[12] Another (unpublished) description by Giovanni da Empoli in 1514 agreed that the Chinese were "white men [*uomini bianchi*], dressed after our fashion like Germans, with French boots and shoes." The earliest reports of the Japanese took much the same form, although at first information was available about Ryukyu islanders only. Pires had noted that they were "white men, well dressed" (*homees bramquos bem vestidos*), but they were also said to be even better than the Chinese, "more dignified." This is because they were perceived to be wealthier, and in 1517 an order was sent out to find the location of these people of "Lequia."[13]

My review of these early accounts is both cursory and oversimplified, but it should be clear immediately that the whiteness or reputed whiteness of these people was not a racial designator and indeed not really even a description of their color. If the Chinese or Okinawans were described as white, it was a function of their affluence and their power and their apparent level of cultural sophistication. White, like all color terms, was evaluative rather than descriptive. Perhaps East Asian pigmentation was seen to differ from that of Africans or Indians or Malaysians, but this was not why they were called "white." Before long it was also a function of their perceived capacity to become truly "civilized," which is to say converted to European Christianity. It is partly for this reason that the Japanese started out much whiter than their Chinese neighbors, since by the end of the sixteenth century hundreds of thousands of people had already been converted. The published version of the first European visit to Japanese territory in 1543, when a group of Portuguese had landed by mistake on the island of Tanegashima, noted that the boats sent out to meet them contained men whiter than the Chinese (*mais alvos que os Chins*), with small eyes and short beards. Other early reports, such as that of Portuguese captain Jorge Alvares in 1547, agreed that these were a white people (*gemte bramqua*), and the same phrase was used much more famously in the 1552 "Letter to Europe" of St. Francis Xavier, the pioneer missionary who had arrived in 1549.[14]

Henceforth, Japanese whiteness would be constantly reiterated in the missionary literature, but this, too, was always connected to the (potential) Christianity of the people. Xavier notoriously described them as "the best people yet discovered" ("among the infidels," that is), but this was a verdict thoroughly based on local codes of honor, honesty, and virtue that were perceived as being amenable to Western Christianity. As fellow missionary Balthasar Gago put it in 1555, despite the fact that the Japanese were a handsome, polite, and cultured white people (*gente branca*), it was no marvel to see them fall into misery, as they lacked the true religion. Or as another (anonymous) missionary wrote in 1581, the Japanese might well be white (*colore candidi*) and even superior to many Europeans, but it was precisely those qualities that made them "especially fit for taking up the religion of Christ."[15]

Not everyone would share this sort of enthusiasm, however, and many insisted that the Japanese were colors other than white. Alessandro Valignano, for instance, the Jesuit Visitor to Japan and chief architect of the program for conversions, disgustedly noted how many Portuguese continued to refer to the Japanese (and the Chinese as well) as *negros*. Other nonmissionary eyewitness travelers, such as Olivier van Noort and Dirck Pomp, found the Japanese brown (*bruyn*) and black (*swart*) instead.[16] And when four Japanese teenagers made a papal visit to Rome in 1585, a public relations event carefully orchestrated by Valignano as well, they were variously described as olive (*olivastro*), brown (*bruna*), pale and deathly (*pallida e smorticcia*), the color of Africans (*di colore Affricano*), and the color of lead (*di color di piombo*). And this was despite the fact that they had already been converted. One commentator puzzled over the fact that as they lived in a cold climate, they "should be white" (*suol esser bianca*), and he tried to explain this fact by appealing to the idea that perhaps their olive (*olivastro*) skin was simply a result of their long and arduous journey.[17]

Soon enough, however, Japanese Christians would become the object of increasingly frequent and often horrific persecutions. In 1614 the religion was officially banned, and in 1639 the last of the remaining Westerners had been summarily expelled. The tone of all European reports, as one might imagine, grew steadily more subdued along with the complexion of the natives. When an actual Japanese ambassador showed up in Rome in 1614, he was judged to be pallid (*palles*), tanned (*basannéz*), yellow tending to olive (*giallo, tirante quasi all'ollivastro*), and greasy black (*di color nero grasso*).[18] In 1660 even the official Jesuit history now called them olive (*ulivigno*). Another Jesuit text from 1715 continued to assert that the Japanese were at least less olive (*moins olivâtre*) than other East Asians, but this view was revised twenty years later when the same author proclaimed that their color was just as olive (*olivâtre*) as the Chinese.[19] In the meantime Engelbert Kaempfer's *History of Japan* had been published in 1727, which remained Europe's standard account until well into the nineteenth century. Here the Japanese were no longer white but brown (*braune*) instead, and this was rendered into English as tawny and into French as *bazanez*, which along with *olivastro* were perhaps the most common terms for describing native peoples who were not exactly black but not white either.[20]

A PLETHORA OF COLOR TERMS

China, on the other hand, remained much more intractable to European religious and mercantile aspirations from the start. Xavier died in 1552 without having realized his ultimate dream of converting the Chinese, whom he described as both white and even more intelligent than the Japanese. Chinese whiteness and aptitude were similarly emphasized in a description of the empire credited to Valignano and published in 1605 (and then again in *Purchas His Pilgrimes* in 1625), as well as in such texts as Orazio Torsellino's life of Xavier of 1594.[21] By the close of the sixteenth century Xavier's successor, Matteo Ricci, was finally able to establish a permanent missionary presence in the country, inaugurating a policy of accommodation in which conversion had to be accomplished from within by mastering and appealing to local traditions and languages as much as possible. But by this time Chinese whiteness was already being routinely qualified by allusions to considerable tonal variation in different parts of the country. Thus while Ricci stressed that the Chinese were white (*albi*), he also noted that people in the south were naturally darker (*subfusci*) than those in the north.[22]

While this might appear to be a simple appeal to standard theories in which skin color is determined by climate (and there were similar statements regarding northern and southern Japan), I would argue that the insistence on the darkness of (some of) the Chinese was inextricably related to the difficulty of Christianizing them. The Japanese mission may have been a fantastic (if meteoric) success, but things were different in China, where the paucity of conversions was an embarrassment for the missionaries and was always blamed on the Chinese themselves. Ricci complained in 1596 that in the southern province of Guangdong only one hundred conversions had been achieved in more than fifteen years. A not unrelated stereotype, which began at exactly the same time, was that Chinese children were born white and beautiful but that they became uglier, fatter, and darker as they grew older.[23]

A few missionaries continued to insist that Chinese skin was unqualifiedly white. One was Martino Martini in 1655 in his influential Chinese atlas. Another was Alonso Sánchez, who in 1587 had argued that as the Chinese where white (*blanca*) and "had nothing Indian about them," they would gladly throw off their despotic form of government, intermarry, and ultimately accept the protection of the Spanish crown. This led him to propose a wild scheme to colonize the country with only a few ships and a few thousand armed men, in imitation of earlier Spanish "successes" in Mexico or the Philippines.[24] Yet it was far more common to find the Chinese differentiated according to some combination of geography, gender, age, and social rank. While this may have seemed more sensible, it also produced great divergence of opinion. By 1600 published reports had described the Chinese as brown, red, tawny, black, and swarthy, and over the next two hundred years they were also dun, tanned, ashy, sanguine, olive, brunette, ruddy, and florid—and these were only the English terms.[25]

Indeed this lack of consensus multiplies exponentially when one considers this wealth of material from a more international perspective, as the same color terms become translated and retranslated into different cultural and linguistic traditions. I can think of no better example than the most influential book about China of the entire early modern period, Juan González de Mendoza's *Historia... del gran reyno de la China* of 1585, which over the next century would appear in more than fifty editions in Spanish, Italian, French, English, and Latin. Mendoza, an Augustinian monk, was instrumental in disseminating the idea that the true color of the Chinese covered a wide spectrum, owing to the country's immense size and its varied climate. This information would soon appear in books on China of every imaginable kind: from atlases to missionary accounts to ambassadorial and other travel narratives.[26] Yet Mendoza's book was part of a larger rhetorical strategy to counteract previous reports that the Chinese were dark. In Canton, he said, which is adjacent to Maca and thus the only region that most Europeans would know anything about, people were brown (*morenos*) as in Fez or in Barbary, because they lived in the same parallel. But those of the interior were white (*blancos*), some more than others, according to their proximity to a colder climate. Some of these people, Mendoza concluded, resembled the Spaniards in terms of color, and others were still more fair (*rubios*), even approaching the fair (*rubios*) and red (*colorados*) coloring of Germans.[27]

It is not always easy to pinpoint exactly what kind of colors Mendoza was trying to delineate, particularly the term *rubio*, which could mean either fair or blonde. To make matters worse, when he returned to the same question in a later chapter the color terms multiplied. Those from Canton were said to be Moorish (*amoriscados*), while others further north were the color of Germans, Italians, and Spanish: white (*blancos*) and fair (*rubios*) or a bit dark green (*verdinegros*)![28] An Italian translation the following year said that the Cantonese were olive (*olivastri*) instead of brown, and *rubio* was translated as blonde (*biondi*). *Verdinegro* was also rendered as olive (*olivastri*), thus creating a confusion in which both the darker southern regions and the supposedly paler ones further north were similarly olive-skinned. A Dutch version of 1595, translated from the Italian, presented the same view, with both the Cantonese and inlanders being olive (*olyvich, olyachtich*), even if the latter were also potentially blonde (*blont*). The French text made the Cantonese black (*noiraux*), while those of other regions were said to resemble blonde (*blonds*) and red (*rouges*) Germans. *Verdinegro*, as always the most challenging term, was doubled as both greenish brown (*verdbruns*) and tanned (*basanez*). The English version was even more problematic, since while it agreed that the Cantonese were "browne people," those in the north were said to be "more yealow, like unto the Almans [i.e., Germans], yelow and red colour" (figure 3).[29]

If we can for the moment distance ourselves from this apparently sudden appearance of the term "yellow" (it is not without precedent; I have intentionally omitted it from our survey thus far), let us note that the English translator has rendered both *blanco* and *rubio* as yellow, and that this was meant to be a positive rather than a negative term. If some of the Chinese were yellow, in other words, it was because they were fair-skinned like northern Europeans, not because they were what we would now call "people of color." Yet in the English version the reoccurrence of *rubio* in Mendoza's later chapter was translated as "red" instead of yellow, probably based on a simple mistranslation in which *rubio* (fair) and *rubí* (red) were being elided or confused

CHAP. II.

Of the temperature of the kingdome of China.

The temperature of this mightie kingdome is diuersly, by reason that almost the whole bignesse therof is from the south to the north, in so great a length that the iland of Aynan being neere vnto this land, in 19 degrees of altitude, have notice of some prouinces that are in more than 50 degrees, and yet they do vnderstand that beyond that there bee more vpon the confines of Tartaria. It is a strange thing to be seene, the strange and great difference betwixt the colours of the dwellers of this kingdome. In Canton, a mightie citie, whereas the Portingales had ordinarie trafficke with them of China, for that it was nigh vnto Macao, where as they had inhabited long since, and from whence they do bring all such merchandise as is brought into Europe. There is seene great diuersities in the colours of such people as doe come thither to trafficke, as the said Portingales do testifie.

Those which are borne in the citie of Canton, and in al that cost, are browne people, like vnto them in the citie of Fez or Barbarie, for that all the whole countrie is in the said paralel that Barberie is in. And they of the most prouinces inwards are white people, some more whiter than others, as they draw into the cold countrie. Some are like vnto Spanyards, and others more yealow, like vnto the Almans,[1] yelow and red colour.

Finally, in all this mightie kingdome, to speake generally, they cannot say that there is much cold or much heat, for that the geographers do conclude and say it is temperate, and is vnder a temperate clime, as is Italy or other temperate countries, wherby may be vnderstood the fertilitie of the same, which is (without doubt) the fertilest in all the world,

[1] Germans.

Figure 3: "Of the temperature of the kingdome of China," from Juan González de Mendoza, *The History of the Great and Mighty Kingdom of China* (1853–54 ed.). This translation originally appeared in 1588. "Those which are borne in the citie of Canton . . . are browne people, like unto them in the citie of Fez or Barbarie. . . . And they of the most provinces inwards are white people, some more whiter than others, as they draw into the cold countrie. Some are like unto Spanyards, and others more yealow, like unto the Almans [Germans], yelow and red colour." Princeton University Library.

(although they are related words). Mendoza expressed red as *colorado*, not *rubí*, and in the earlier chapter the English translator did get it right.[30] *Verdinegro* was "somewhat swart." Finally, both the German and Latin editions (they were produced by the same publisher) described the Chinese as yellow (*gelblichter*; *subflavescunt*), but this time the color was applied to the darker Cantonese rather than the German-like northerners. But lest we jump to the conclusion that yellow had now become a "colored" category in the modern sense of the term, both the German and Latin texts also utilized yellow to describe the more fair-skinned residents of the north, who were white (*weißfärbig*; *albicant*) and pale (*bleich*; *pallent*) and yellow (*gelblicht*; *subflavescunt*) like the Germans. *Verdinegro*, perhaps unsurprisingly, has been silently omitted.[31]

NO LONGER WHITE

What did it mean, then, if the Chinese were called yellow or *gelblich* or *flavus*, a decidedly ambiguous term from the start? Depending on the language in which in which it was expressed, yellow could carry either positive ("fair," "blonde") or negative ("colored," "nonwhite") connotations. Or perhaps we should say that yellow began as a way of emphasizing Chinese proximity to Europeans ("not all Chinese are dark"), but that over time it had become redeployed as a term of complexional distance ("no Chinese is white")? Similarly, a number of early travelers to North and South America described the natives they encountered as fair-skinned and indeed white (it was also hypothesized that Native Americans were descendants of the Chinese). But Native Americans slowly became categorized as "redskins" at exactly the same that East Asians began to be yellow, and analogous patterns could be said to exist in other regions visited by Europeans.[32]

In 1555 Francisco López de Gómara's *Historia general de las Indias* marveled at the different skin tones in the Americas, which he said varied between russet (*bermejo*), fair or blonde (*ruvio*), ashy (*cenizoso*), brown (*moreno*), tawny (*leonados*), the color of cooked quince (*membrillos cochos*), yellow/jaundiced (*tiriciados*), and chestnut (*castaños*). This, too, was a classic piece of terminological disorder, compounded by Richard Eden's English rendering of *ruvio* as yellow, *moreno* as "murrey" (i.e., mulberry), *leonado* as purple or tawny, and *tiriciado* as olive.[33] Yet as for the term "redskin," I have mentioned in the introduction that there were early reports of people in war paint or anointing themselves with plant substances that might indeed have given their bodies a reddish tinge. Yellow was another of these "intermediate" shades that was neither white nor black, although it had no such "real" excuse like Indian redness to fall back on.[34] In the Chinese context yellow clothing was reserved for the use of the royal family, one of the mythical founders of China was known as the Yellow Emperor, and the Yellow River (the central color, the color of the earth) was a vital feature of many aspects of Chinese cultural life. But none of these things could hope to explain why East Asian (and not just Chinese) skin should be yellow, too.

Europeans often appealed to food or other natural phenomena to explain the colors of the people they encountered (and thus the early comparisons between Native Americans and quinces, chestnuts, mulberries, or ashes). At the end of the eighteenth century Johann Friedrich Blumenbach would place East Asian skin halfway between the color of cooked oranges and grains of wheat. But it is noteworthy that East Asians were just as frequently compared to dried-up or lifeless things, in Blumenbach as well, who mentioned desiccated lemon peel.[35] One of the earliest published references to the yellow Japanese, in Arnoldus Montanus's record of two mid-seventeenth-century Dutch trade missions, published in 1669, had a similar valence. It was here claimed that while in relation to other East Indians the Japanese may be white (*blank*), they were both yellowish (*geelachtigh*) and lacking a lively color (*zonder levendige verve*) when compared to Europeans. As the French translator even more chillingly put it, the Japanese were *d'une couleur morte*.[36] Similarly, the French translator of Giovanni Pietro Maffei's popular *Historiarum indicarum* of 1588 noted that the southern Chinese were *livide*, even though they were bronze (*aeneo*) in Latin and *olivastri* in Italian. In the 1610s John Saris described a group of upper-class Japanese women as "cleare skind and white, but wanting colour." Francis Bacon remarked that the Chinese were "of an ill complexion (being olivaster)." And in 1741 a German Jesuit missionary wrote that even though the Chinese nobility might indeed appear white (*zwar weiß*), it was a whiteness that was more of a deathly pallor (*eine Todten-Bleiche*).[37]

For Carl Linnaeus, finally, who will play a key role in the newly developing fields of natural science analyzed in the next chapter, the term *luridus* (sallow, wan, ghastly), the color he eventually chose to classify *Homo asiaticus*, was more of a sickly yellow than a golden one. In both botany and medicine, his real areas of expertise, it was the color of disease, and this was almost certainly linked to a new eighteenth-century sinophobia that saw the Chinese no longer as white, civilized, morally superior, and capable of Christian conversion, but instead as pale yellow, despotic, stagnant, and forever mired in pagan superstition.[38]

Yellow could undoubtedly carry such negative connotations from the beginning, but it is a historical mistake to claim that in the early modern period the term appeared with any consistency or consensus, or that its appearance could somehow explain a fully racialized example like the late-nineteenth-century phrase "the yellow peril." In 1770 a Chinese visitor to London was described as having a "complexion different from any Eastern[er] I ever saw, with more yellow in it than the Negroes or Moors"; but a competing account in the *Gentleman's Magazine* said he was "of a copperish colour."[39] Another interesting example appeared in an exchange of letters in 1785 between George Washington and his former aide-de-camp Tench Tilghman. When describing a small group of Chinese sailors working on a cargo ship, Tilghman noted that they "are exactly [like] the Indians of North America, in Colour, Feature, Hair, and every external Mark." "Before your letter was received," Washington replied, "I had conceived an idea that the Chinese, tho' droll in shape and appearance, were yet white." No, said Tilghman, still repeating Mendoza after two hundred years, "the Chinese of the Northern provinces are fairer than those of the south, but none of them are of the European Complexion."[40]

Ultimately, in the case of Chinese people, beneath the confusion about which term seemed to describe their complexions best—yellow, tawny, brown, black, red, copper, or dark green—they were not white, or at least not white any longer. A further example was provided by an early-seventeenth-century record of the journeys of an Austrian traveler named Christoph Carl Fernberger, which remained unpublished until 1972. During his visit to Quanzhou in the southern province of Fujian, which he found to be "the worst place in China," the natives were a little bit yellow (*ein wenig gelb*), even though they were also well proportioned. Fernberger also condemned them for pederasty and for the fact that old people were forced to do menial tasks to earn money. In the extreme southwestern Japanese city of Hirado, however, where there were still many Christians, the people were correspondingly described as rather white (*zimblich weiß*), even if they were smaller than Europeans. In contrast to the Chinese, moreover, these people lived well and were a "manly nation in war" (the Chinese were "effeminate"), and their women were white and charming (*weiß und anmütig*).[41] Clearly implicated in these color terms (and this is one of our earliest examples of a traveler who actually refers to the Chinese as yellow) were complex and multiple distinctions based on perceived levels of culture, religion, morality, stature, social rank, and gender, among others. And it would not be long before the Japanese, too, lost their whiteness and became just as yellow (or olive) as their neighbors in China.

WHY YELLOW?

We might return for a moment to the case of the Japanese teenagers in 1585, since they represent a rare early case of East Asian bodies actually being described in a Western context (there were others, but this instance is one of the most obsessively documented). Although they were merely guests and not immigrants, the overwhelmingly negative picture we get of their appearance is to be measured against the level of praise they received in terms of their comforting Christianity. Their politeness and their noble bearing were constantly reiterated, but so too were their small "protruding" eyes, their beardlessness, and their compact stature. More than one eyewitness added that they looked alike.[42] Their arrival may not have been the subject of a peril but it did set off something of an East Asian fever in late sixteenth-century Europe. Mendoza's book about China came out that same year.

Among nearly one hundred accounts of the embassy that have come down to us, one of the most interesting is a heavily fictionalized Latin dialogue published in 1590 that purported to record the boys' discussions about their recent experiences in Europe. Produced under Valignano's direct supervision, the text was hardly told from a Japanese perspective. This is nowhere better exemplified than in the book's climactic final colloquy, where Western preoccupations with skin color and its relationship to culture were given full vent. One of the boys was made to remark that while East Asians might live in the same parallel as Europeans, they were not really equal. The Chinese or Japanese (or Persians or Arabians) might have been endowed with intelligence and a white color (*albo colore*), but they were not fully versed in the humanistic and liberal disciplines (*humanitatem & liberalem disciplinam*)—in other words, the Western Christian tradition.

But could a non-European ever be fully versed in this tradition? Judging by Valignano's other writings it might seem that the Japanese were indeed capable of it, their whiteness even according them the unique honor of being considered candidates for the Jesuit order. For Valignano, in fact, the idea of Japanese whiteness was absolutely essential and distinguished them from other Asians—especially the residents of India, whom he, too, called *negros*. And yet the 1590 colloquy still concluded that while some non-Europeans might be fair and Christian and illustrious men (and of course the Japanese boys themselves had already been converted), what distinguished Europe from every other region of the world was that it had been blessed by God as the center of human civilization.[43] Although the precise relationship between that version of civilization, Christianity, and fairness of skin was never made explicit, the implications were clear: even a civilized and converted East Asian could never become truly white. Much of this was also summed up by a 1698 description of Chinese residents in the Indian city of Goa, "who are White, Platter-fac'd, and Little-eyed, tolerated on account of embracing Christianity."[44]

It had become necessary to ensure that Asians were safely distanced from a whiteness that only the West was allowed to embody—and a whiteness that was beginning to be defined at exactly the same time. Writing from Beijing in 1602, Jesuit Diego de Pantoja noted that even though the Chinese were completely white (*todos blancos*) they were not so white as the Europeans. As Pantoja's English translator put it, "they are commonly all white, yet not so white as those of *Europe*: and therefore to them we seeme very white."[45] Much as in the examples of Sernigi and Camões with which we began, there was an obvious assumption here that only Western Europeans could be "very white," and moreover that this lack of color was a sign of the West's intellectual, cultural, and religious superiority. And yet what Pantoja was evidently unable to see was that in East Asia whiteness was not necessarily a positive term at all. If the Chinese considered Westerners white, it may well have been to associate them with the color of death; these were "white devils" rather than superior light-skinned bearers of civilization and Christian salvation.

Note, too, how Mendoza's hierarchy of colors has been inverted in Pantoja's statement, since according to Mendoza's point of view, if some of the Chinese were fair-skinned, it proved that this was a superior culture in which the people were not, as was stereotypically supposed, "tinged with color." But in Pantoja's account, and almost exclusively thereafter, the whiteness of East Asians was ultimately relativized. Chinese and Japanese might be "white," but Westerners were even whiter, and as the people of East Asia darkened, their once varied skin tone gradually ossified into a perceived sameness ("they all look alike").[46] Eventually, despite regional or class or gender differences *all* of them were yellow, although some were lighter than others.

In the case of China, even the article included in the famous 1910–11 edition of the *Encyclopaedia Britannica* agreed that while their complexions varied "from an almost pale-yellow to a dark-brown, without any red or ruddy tinge," "yellow, however, predominates."[47] But this was a trend that had been long in the making. In 1740 Dominique Parrenin, a Jesuit missionary working in Beijing, wrote that no people were more uniform or more homogeneous, although their color did vary from *blanc* to *basané*. David Hume's essay "Of National Characters," first published in 1748, noted that "the *Chinese* have the greatest Uniformity of Character imaginable" (owing to the country's "extensive government"), and in his *Description générale de la Chine* of 1787, French Jesuit Jean-Baptiste Grosier added that all Chinese seemed to be cast from the same mold. In none of these treatments, we should note, were the Chinese specifically called yellow.

But by the time of the Protestant missionary Karl Gützlaff's *Sketch of Chinese History*, published in 1834, the yellow stereotype seemed complete. "The Chinese features are, in themselves, not very handsome: a small eye, and flat nose, a yellow complexion, and a want of expression in the whole countenance." "It is truly astonishing," Gützlaff continued, "that in so extensive a country as China, . . . no greater variety in the human race should be found. Not only is there the greatest sameness in the color of the eye and the shade of the hair, but the inhabitants of the various provinces differ very little in their whole outward appearance. Nor is this characteristic sameness confined to the body, it extends also to the mind."[48]

"Yellowness" had become a feature of the Chinese mind as well as the body; it was no longer just a color. Much the same was true for descriptions of Japan, although there was far less information available during the two centuries in which it was largely closed to the outside world beginning around 1640, with the exception of a Dutch-operated trading post on the island of Deshima in Nagasaki Bay. This lack of information also enabled an impostor like George Psalmanazar, about whom I have written elsewhere, to pose in 1703 as a native of Formosa (modern Taiwan), which he said was an island subject to Japan. As far as we know, Psalmanazar was a white and blonde Frenchman (his true identity has never been discovered), and while the question of skin color did come up it was never a serious hindrance to his disguise, partly because of conflicting knowledge about the island and partly because the very concept of coloring had not yet been fully fixed as a racial marker. Eyewitness accounts of Formosan skin tone varied considerably as well: the men were described as between black and brown (*tusschen swart ende bruyn*), blackish yellow (*zwartachtig geel*), "of a dark brown complexion," and "tawny or olivaster"; and the women were between brown and yellow (*tusschen bruyn en geel*), yellow, "inclining more to an olive colour," and "bright like the golden yellow."[49] Psalmanazar's reply was that the upper classes in Formosa (to which he belonged) lived mostly underground and thus were not exposed to the sun, an explanation perfectly in keeping with contemporary notions about the climatological nature of complexional differences.

But for the most part, as we have mentioned above, the Japanese were described as brown or olive-skinned throughout the eighteenth century, although they also began to be seen as yellow—and perhaps earlier than the Chinese. Beginning in the 1770s, Carl Peter Thunberg, a student of Linnaeus and almost certainly affected by the master's classification of Asians as *luridus*, testified on numerous occasions that the Japanese were *gulagtig*, rendered in other languages as *gelblich*, *jaunâtre*, and yellowish. A second French edition of his travels, however, preferred to translate the Swedish word as *basané* and *cuivré*. Thunberg also emphasized that upper-class women were completely white (*fulkomligen hvita*), and little new material was obtainable until Philipp Franz von Siebold, a physician and naturalist who like Kaempfer had been in the employ of the Dutch at Deshima, began to publish his *Nippon* in serial form in 1832. Undoubtedly influenced by another great taxonomist, Siebold repeated Blumenbach's categorization of Asians as *weizengelb* or wheaten-yellow, while the higher ranks in Japan were white (*weissen*) and beggars varied between copper-red and earth-colored (*Kupferroten und Erdfahlen*). Charles MacFarlane's *Japan* of 1852 agreed that Japanese faces were yellowish.[50]

Just a year later, however, the expedition of Commodore Matthew Perry of the U.S. Navy would force Japan to end its long-standing policy of self-seclusion, and the tone of Western descriptions briefly, and only occasionally, changed once again. Japan was now seen as an enchanting and exotic wonderland of cleanliness, order, and diminutive cuteness, and especially in contrast to China, whose international reputation had fallen dramatically since the middle of the eighteenth century. The Japanese now needed to be carefully distanced from nearby China, as in the ambassadorial chronicle of the marquis de Moges in 1857–58, in James Lawrence's voyage narrative of 1870, or in Charles Eden's popular compilation published in 1877. All of these authors insisted that the Japanese were not yellow like the Chinese; Eden said that they were swarthy, copper-colored, and olive brown instead, while de Moges raved that they were "as white as we are" (*aussi blancs que nous*) in a shadowy reenactment of the earliest European reports from the sixteenth century. According to Lawrence, "many of them would pass for Europeans." Yet the power of Western racialized thinking had become so strong that neither Chinese nor Japanese faces could really be seen as white once more. Both nations had become fully categorized as members of a "yellow race"; as early as 1813 the English ethnologist and one of the first to adopt Blumenbach's five-race scheme, James Cowles Prichard, noted that while there might be frequent fair-skinned variations of the "Mongol" type, which embodied "a yellowish tint passing into an olive," this was only because their color had become "softened and mitigated."[51]

In sum, both China and Japan lightened and darkened depending on thoroughly Western prejudices and Western preconceptions, but this still does not answer the question: why yellow? We have begun to trace the history of that particular term in travel accounts and in missionary texts, and we have seen a clear although often confused and fitful movement from whiteness to yellowness. But how and why did Westerners see East Asians as yellow in the first place? In order to understand that particular way of seeing, we need to look elsewhere. We can, I think, begin by examining the taxonomical sources cited by such eyewitnesses as Thunberg and Siebold, since what lay behind their appeals was a new and different conception of racial difference that had come not from earlier travel authorities or missionary precursors, but from the world of natural science. Here, perhaps, a better source for yellowness might be found, and this is the subject of chapter 2.

Chapter 2

Taxonomies of Yellow

*Linnaeus, Blumenbach, and the Making of a
"Mongolian" Race in the Eighteenth Century*

In the last chapter we have begun to see how Western descriptions of East Asian people fitfully moved from calling them white to calling them yellow, although it was ultimately unclear why yellow should have been chosen from among so many other possibilities, including tawny, *moreno, olivastro, basané, gefärbt, fuscus,* and *bruin*. I have suggested that we are not going to find a satisfactory answer to that question by looking at travel texts or other forms of "eyewitness" description, simply because the adoption of any color term was symptomatic of a larger development within racialized thinking itself. To call East Asians yellow, in other words, was a means of ensuring that while they might not be as dark-skinned as Africans, they could no longer be considered "white" either.

To some degree, perhaps, a new color had to be found for them. By the middle of the eighteenth century all human groups began to be pigeonholed according to new scientific conceptions of heritage and biological descent, and it is from the world of natural science that a more satisfying answer to yellowness might be found. Older climatological theories about skin color, dating back at least to Aristotle, had numerous difficulties to overcome. How was it possible to explain the markedly divergent appearance of people who lived in similar climates, or that people did not really change color simply because they had relocated to another environment? Jean Bodin's influential *Six livres de la république* (1576) represents an important example, as it included a long chapter outlining an environmental scheme in which both the northern and southern hemispheres of the world were divided into three regions—hot in the tropics, cold near the Poles, and a temperate zone in between. People who lived to the north and south were distinguished by their relatively whiter or blacker skin, and Bodin conjectured that there were cultural differences between east and west as well: the Chinese, the easternmost people, were the most civilized ("the most ingenious and courteous people in the world"), while Brazilians, the furthest west, were the most barbarous.

Humoral theory was also invoked. While middle peoples were sanguine and choleric, southern peoples were more choleric and melancholy, and thus more "sunburned with black and yellow" (*basannez de noir & de jaune*). These contentions were based on a familiar combination of travelers' reports and contemporary cliché, and other popular collections such as Joannes Boemus's *Omnium gentium mores* (1520) or Sebastian Münster's *Cosmographia* (1544) had maintained broadly similar geographical arrangements. But these texts, too, had to rely on standard ancient and medieval authorities such as Herodotus, Pliny, and Marco Polo when they came to discuss Asia, a continent that continued to feature a variety of fabulous and monstrous peoples that were duly enumerated as well.[1]

During the eighteenth century climatological explanations were being slowly abandoned, even though the new human taxonomies that began to take their place—by Linnaues, Buffon, and Blumenbach, in Sweden, France, and Germany, respectively—continued to be heavily influenced by the same stockpile of information about the world's peoples. Linnaeus, Buffon, and Blumenbach each attempted to categorize the world according to (among other factors) the color of the skin. But I would also argue that it was actually the placement of *Asia*—and eventually East Asia—that presented an especially challenging difficulty for these naturalists, since unlike the supposedly clear and historically validated distinctions between the "white" and "black" races, the other peoples of the world had to be colorized without clearly defined precedents. "Red" Americans also presented something of the same dilemma, but Asia was much larger and more diverse and was not a continent of supposedly cultureless savages living in a perceived state of nature. What single color could be found to describe Asia, "the East," "India"?[2]

YELLOW INDIA

If there was any sort of stereotype about "yellow people" in travel texts published before the second half of the eighteenth century, in fact, it did not refer to the people of China and Japan. Rather, yellow could be identified with Jews or with Semitic peoples in general, as in the readings of the tomb of Seti I that we examined in the introduction. Yellow could also be associated with Native Americans, as when (in 1590) Giordano Bruno noted that human beings came in many colors including black Ethiopians and yellow (*fulva*) Americans. Sometimes yellow was used for northern Africans, too: the term "tawny moor" (as opposed to "blackamoor") is yellow moor in Dutch and French. Finally, and perhaps most often with regard to Asia, Europeans used yellow to describe people in India.[3]

This was certainly the case in the Western text that is typically credited with being the first to propose an organization of the world according to race, although the term was used much more vaguely than in the modern sense. A mere seven pages long, it first appeared anonymously in the *Journal des sçavans* in 1684, one of the premier scientific and literary periodicals of its day. Titled "New Division of the Earth, According to the Different Species or Races of Man that Inhabit it, Sent by a Famous Voyager," it was composed by François Bernier, a physician, traveler, and philosopher who among his other experiences had spent twelve years as the physician of a high official in the Mughal court in India.[4]

Bernier's text was deceptive in its brevity and its seemingly casual style, but it was also quite serious in intent and did not go unnoticed at the time. Leibniz, for instance, who disagreed with its premise, referred to it in a letter of 1697 that was subsequently published (along with an excerpt from Bernier translated into Latin) in 1718. "Up to now," Bernier began, "geographers have divided the world according to the different nations or regions they have found."

> But what I have noticed in my long and frequent voyages has given me the idea of dividing the world in a different way. Although in the exterior form of their bodies, and principally in their faces, nearly all

people differ from each other according to the different districts of the Earth that they inhabit (in such a way that those who have often traveled are not normally mistaken in their ability to distinguish each nation in particular), I have nevertheless noticed that there are four or five Species or Races [*Especes ou Races*] of men whose difference is so remarkable that it may serve as a just foundation for a new division of the Earth.

What was new in this "new division" was that it attempted to distinguish people according to their "exterior form" rather than their cultural diversity, and Leibniz, who was at the time more interested in dividing the world according to language, objected precisely to this point. Variations in "size and constitution of the body," he countered, do not "prevent all human beings who inhabit the Earth from being of the same race [*d'une meme race*], which has been altered by different climates just as we see animals and plants changing their nature and becoming better or degenerating."[5]

But for Bernier these differences were more elementary and could not be explained simply by the effects of environment. The first "race," he said, consisted of Europeans, northern Africans, and people of the Middle East, India, and parts of Southeast Asia (a broad grouping that does not fit very comfortably with modern prejudices). Among this group the Egyptians and Indians were *fort noirs* or *bazanez*, but this was due merely to an increase in their exposure to the sun. Those of the upper classes were no darker than the Spanish. While many people of India, finally, "have something very different from us in the shape of their faces and in their color, which often tends toward yellow" (*jaune*), this was not enough to make them a separate species, just as one would not want to separate the Spanish from the Germans on similar grounds.[6]

If any people were "yellow" in Bernier's presentation it was certain inhabitants of India and not East Asia, and a similar verdict appeared when he waxed poetic on the beauty of Indian women, whose color "tends just a little bit to yellow" (*jaune*). While I am unable to provide a full catalogue of seventeenth- and eighteenth-century references to "yellow Indians," and especially Indian women, examples include

Montesquieu's best-selling *Lettres persanes* (1721), where women from the kingdom of Vijayapur were called *jaunes*; Voltaire's *Traité de métaphysique* (1734), in which, clearly in the context of India, it was said that those of the white race could never produce *peuples jaunes*; and Nicolle de la Croix's *Géographie moderne* (1752), which explicitly referred to Indians as *jaunâtres*. Finally, in Immanuel Kant's brief treatments of race from the 1770s and the 1780s Indians were repeatedly described as *gelb* or *olivengelb*.[7]

Interestingly, Bernier did not actually refer to his "first species" as white (although it was certainly implied), and he offered no detailed information about their appearance either. This was presumably because it was only the other races that needed to be differentiated from the "normal" one. The "second race," however, was explicitly identified as black, thick-lipped, beardless, oily-skinned, and wooly-haired, and this blackness, he continued, was not caused by climate but by "the peculiar texture of their bodies, their semen, or their blood." The "third species" consisted of what we now call East Asia, as well as most of southeast Asia, Tartary, and "a small part of Moscovy." These people were *véritablement blanc*, a verdict much in accord with other period descriptions. The difference of these people, rather, lay in their "broad shoulders, flat faces, small hidden noses [*petit nez écaché*], little, long, and deep-set pig's eyes [*petits yeux de porc, long & enfoncez*], and three hairs of a beard."[8]

As in the case of the "African race," Bernier's unpleasant racial stereotyping is apparent, but it was equally important that the contours of his divisions and their accompanying colors did not yet match what would become the accepted racial categories by the end of the nineteenth century. Similarly, Bernier's "fourth species" was a rather unusual choice, since he singled out the far northern people of Lapland as a separate race, who were "very hideous and look much like bears." Lastly, he mentioned the *olivastres* Americans, who, despite the fact that their faces were "turned out in a different way from ours," might not actually be a race unto themselves.

At this point the text lapsed into a salacious Orientalist discussion of Eastern women, including Bernier's own experience of ogling naked women in a slave market. It was also in this context that he specified that yellow Indian women (whom, he adds, "I have found very much to my taste") embodied a bright and vivid yellow rather than an "ugly and livid pallor of jaundice."[9] This is an important reminder that gender was a crucial feature of both colonial travel writing and the development of racialized thought as a whole, and perhaps especially with respect to Asian people, who were both regularly effeminized and objectified as a potential sexual commodity. It was often the case that when a (male) traveler paused to describe native peoples, it was because he was attracted to their relative state of undress or their real or imagined erotic availability (the shameless nudity of "savages," the innocent naiveté of public bathing, and so on). Bernier's description of yellow Indian women was an excellent case in point, as he was careful to add that this skin tone was radiantly attractive and not a marker of ill health. But though Bernier's "new division" was certainly a form of proto-racial categorization and was used by later taxonomers as a precedent for assigning colors to the people of the world, it was once again not exactly a source for the origin of the yellow race, and certainly not with respect to East Asians.[10]

THE FOUR COLORS OF *HOMO SAPIENS*

It is more appropriate to credit that particular piece of imaginative labeling to the Swedish botanist, taxonomer, and physician Carl Linnaeus, although with Linnaeus, too, yellow was not yet fully centered on *East* Asia. Often called the father of modern binomial nomenclature, Linnaeus proposed that the natural world was divided between the animal, vegetable, and mineral kingdoms, and that it could be organized into a single scheme of ever more specific genera and species. He burst onto the international scene with his *Systema naturae* in 1735, which despite its proud and confident title began as a modest volume of only thirteen folio pages (two of which were blank), consisting of three double-page tables accompanied by a few pages of observations and an additional table explaining his theory of the "sexual system" of plants. In his own day Linnaeus was particularly well known for the last of these, a scheme for classifying plants based on the number and arrangement of pistils and stamens, which he referred to as the "nuptials" of plant life. What concerns us here, however, is his treatment of the animal world, and of *Homo sapiens* in particular, a term he also coined.

As the title of his book made clear, Linnaeus's main interest was in *systematizing* the world, which was not merely a question of classification but also of naming. Everything in the world of nature could be placed on a kind of scale of being ("minerals grow; plants grow and live; animals grow, live, and feel"), and *Homo sapiens*, although distinguished as the only creature who "knows himself" (*nosce te ipsum*), was also firmly linked to the rest of the animals—and to primates in particular. This was a modification of the ancient tradition of a great chain of being, in which humans had been placed above the other animals but also below the divine realm of angels and gods. Linnaeus, however, placed the genus *Homo* at the top of the table representing the animal kingdom, and man was further subdivided into four "species," each of which was assigned both a geographical designation as well as a color in a progression from white to black: "Europaeus albesc. / Americanus rubesc. / Asiaticus fuscus. / Africanus nigr." (figure 4).[11]

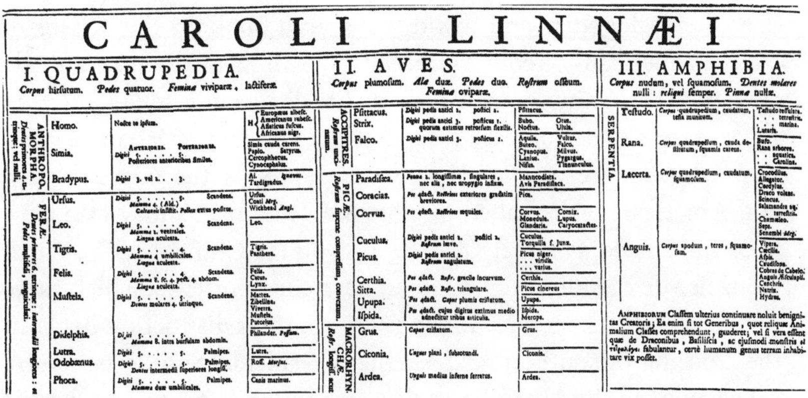
CAROLI LINNÆI

I. QUADRUPEDIA.
Corpus hirsutum. *Pedes* quatuor. *Feminæ* viviparæ, lactiferæ.

ANTHROPOMORPHA. *Dentes* primores 4. utrinque; vel nulli.	Homo.	Nosce te ipsum.	H { Europæus albesc. Americanus rubesc. Asiaticus fuscus. Africanus nigr.
	Simia.	ANTERIORES. POSTERIORES. *Digiti* 5. . . . . . 5. Posteriores anterioribus similes.	Simia cauda carens. Papio. Satyrus. Cercopithecus. Cynocephalus.
	Bradypus.	*Digiti* 3. vel 2. . . 3.	Ai. *Ignavus.* Tardigradus.
FERÆ. *Dentes primores* 6. utrinque: *intermedii* longiores: ... *Pedes* multifidi, unguiculati.	Ursus.	*Digiti* 5. . . . . . 5. Scandens. *Mammæ* 4. (Ald.) *Calcaneis* insistit. *Pollex* extus positus.	Ursus. Coati *Mrg.* Wickhead *Angl.*
	Leo.	*Digiti* 5. . . . . . 4. Scandens. *Mammæ* 2. ventrales. *Lingua* aculeata.	Leo.
	Tigris.	*Digiti* 5. . . . . . 4. Scandens. *Mammæ* 4 umbilicales. *Lingua* aculeata.	Tigris. Panthera.
	Felis.	*Digiti* 5. . . . . . 4. Scandens. *Mammæ* 8. sc. 4. pect. 4. abdom. *Lingua* aculeata.	Felis. Catus. Lynx.
	Mustela.	*Digiti* 5. . . . . . 5. Scandens. *Dentes* molares 4. utrinque.	Martes. Zibellina. Viverra. Mustela. Putorius.
	Didelphis.	*Digiti* 5. . . . . . 5. *Mammæ* 8. intra bursulam abdomin.	Philander. *Possum.*
	Lutra.	*Digiti* 5. . . . . . 5. Palmipes.	Lutra.
	Odobænus.	*Digiti* 5. . . . . . 5. Palmipes. *Dentes* intermedii superiores longiss.	Ross. *Morsus.*
	Phoca.	*Digiti* 5. . . . . . 5. Palmipes. *Mammæ* duæ umbilicales.	Canis marinus.

II. AVES.
Corpus plumosum. *Alæ* duæ. *Pedes* duo. *Rostrum* osseum. *Feminæ* oviparæ.

ACCIPITRES. *Rostrum* uncinatum.	Psittacus.	*Digiti* pedis antici 2. postici 2.	Psittacus.
	Strix.	*Digiti* pedis antici 3. posticus 1. quorum extimus retrorsum flexilis.	Bubo. Otus. Noctua. Ulula.
	Falco.	*Digiti* pedis antici 3. posticus 1.	Aquila. Vultur. Buteo. Falco. Cyanopus. Milvus. Lanius. Pygargus. Nisus. Tinnunculus.
PICÆ. *Rostrum* superne compressum, convexum.	Paradisæa.	*Pennæ* 2. longissimæ, singulares, nec alis, nec uropygio infixæ.	Manucodiata. Avis Paradisæa.
	Coracias.	*Pes* 4dact. *Rectrices* exteriores gradatim breviores.	Pica.
	Corvus.	*Pes* 4dact. *Rectrices* æquales.	Corvus. Cornix. Monedula. Lupus. Glandaria. Caryocatactes.
	Cuculus.	*Digiti* pedis antici 2. postici 2. *Rostrum* læve.	Cuculus. Torquilla f. Junx.
	Picus.	*Digiti* pedis antici 2. *Rostrum* angulatum.	Picus niger. . . . viridis. . . . varius.
	Certhia.	*Pes* 4dact. *Rostr.* gracile incurvum.	Certhia.
	Sitta.	*Pes* 4dact. *Rostr.* triangulare.	Picus cinereus
	Upupa.	*Pes* 4dact. *Caput* plumis cristatum.	Upupa.
	Ispida.	*Pes* 4dact. cujus digitus extimus medio adnectitur tribus articulis.	Ispida. Merops.
MACRORHYNCHÆ. *Rostr.* longiss. acut	Grus.	*Caput* ciliatum.	Grus.
	Ciconia.	*Ungues* plani, subrotundi.	Ciconia.
	Ardea.	*Unguis* medius inferne serratus.	Ardea.

III. AMPHIBIA.
Corpus nudum, vel squamosum. *Dentes molares* nulli: *reliqui* semper. *Pinnæ* nullæ.

SERPENTIA.	Testudo.	*Corpus* quadrupedium, caudatum, testa munitum.	Testudo tessulata. . . . terrestris. . . . marina. Lutaria.
	Rana.	*Corpus* quadrupedium, cauda destitutum, squamis carens.	Bufo. Rana arborea. . . aquatica. . . Carolina.
	Lacerta.	*Corpus* quadrupedium, caudatum, squamosum.	Crocodilus. Allegator. Cordylus. Draco volans. Scincus. Salamandra aq. . . terrestris. Chamæleo. Seps. Senembi *Mrg.*
	Anguis.	*Corpus* apodum, teres, squamosum.	Vipera. Cæcilia. Aspis. Caudisona. Cobras de Cabelo. Anguis Æsculapii. Cenchris. Natrix. Hydrus.

AMPHIBIORUM Classem ulterius continuare noluit benignitas Creatoris; Ea enim si tot Generibus, quot reliquæ Animalium Classes comprehendunt, gauderet; vel si vera essent quæ de Draconibus, Basiliscis, ac ejusmodi monstris οἱ τερατολόγοι fabulantur, certè humanum genus terram inhabitare vix posset.

Figure 4: Table showing the animal kingdom (detail), from Carl Linnaeus, *Systema naturae*, 1st ed. (1735). In the top right corner of the column for *Quadrupedia*, the genus *Homo*, signified by H, is subdivided into colors as "Europaeus albesc. / Americanus rubesc. / Asiaticus fuscus. / Africanus nigr." Princeton University Library.

The Latin participles *albescens* and *rubescens* (whitish, reddish) could be used to suggest a more nuanced value than simply *albus* and *ruber*, although the basic meaning was the same. *Nigr.* was a common abbreviation for *niger*. *Fuscus*, however, was a bit more of a problem since it was so much vaguer. It could certainly be translated as "brown," but it could just as easily mean something loose and baggy like "dark" or "swarthy."

Americans he placed in a separate group. Other authorities such as Bernier had hesitated on precisely this point, and we have briefly mentioned in the previous chapter that there were a number of contemporary theories that Americans were of the same family or "type" as the Chinese. In Bernier's "New Division," moreover, as well as in other texts such as Richard Bradley's *Philosophical Account of the Works of Nature* of 1721, at least some Asians had been included among the "whites."[12]

In Linnaeus, however, all of Asia was called "dark." But why *fuscus*? In classical Latin the term was extremely broad and could refer to darker (Mediterranean) southerners as opposed to paler peoples further north, or to the bronzing effects of the sun, or to Africans.[13] The notion of "dark" or "colored" people residing outside (or on the fringes) of European lands can also be traced in a long line of medieval encyclopedias and world maps. In a book on the origins of Native Americans, Georg Horn called them *fusco*, the same shade as that of the Chinese. In Duarte de Sande's *De missione legatorum Japonensium* of 1590, "Asiatics"—but not Chinese or Japanese, who were white—were described as *fusco* and *subnigro* (i.e., rather black), and Americans were called *fusci* as well.[14] Linnaeus may also have been influenced by widely read reports on the Chinese such as Juan González de Mendoza's *Historia . . . de la China* of 1585, which as we saw in the last chapter had made an important distinction between the coloring of southern and northern residents. Southerners were called *morenos* like the Fez or the Berbers, and this was translated into Latin as *fusci*. Ricci agreed that the residents of Canton were *subfusci* (darkish).[15]

Fuscus, in a word, was exceedingly vague, but while each of the other color terms varied slightly in subsequent editions of the *Systema naturae* (*albesc.* became *albus*; *rubesc.* became *rufescens*; *nigr.* became *niger*), *fuscus* remained constant—although in one exceptional and extremely interesting case, the third edition published in 1740 with German translations of the Latin terms, *fuscus* was given as *gelblich* (i.e., yellowish).[16] This is not an indication that Asians as a group were slowly becoming yellow, however, since it is much more accurate to say that the sudden appearance of this term was yet one more symptom of the long-standing problem of deciding precisely what color "dark" skin was in any particular language. *Gelblich* was simply one among a myriad of popular choices.

FROM *FUSCUS* TO *LURIDUS*

But Linnaeus's taxonomy was to undergo a major alteration in its definitive tenth edition published in 1758–59, as the text slowly mushroomed from a few folio sheets into a two-volume magnum opus totaling more than thirteen hundred pages. Countless corrections and additions had been made over more than two decades, but until the tenth edition the presentation of man had remained largely the same. One small modification was that *Homo europaeus* was no longer given first, although it was not entirely clear why Linnaeus should have adjusted his former hierarchy, particularly since the other major alteration—the addition of a number of descriptive adjectives for each species—was positively suffused with normative claims drawn from the most common contemporary racist stereotypes.

Most importantly, *fuscus* Asians were for the first time called *luridus* instead, which can be variously translated as yellow, pale yellow, sallow, pallid, deathly, and ghastly.[17] While it is frequently claimed that this is the true origin of the idea of a yellow race, it is a suggestion that needs to be treated with considerable care. In the first place Linnaeus did not choose much more common or more neutral terms such as *flavus*, *fulvus*, or *gilvus* (and there were others, too); in classical Latin *luridus* was often pejorative, usually with connotations of horror and ugliness and pallor as in the English word "lurid."[18] Indeed, the choice of *luridus* has not really been seriously examined, although many readers have duly noted that the reasons for Linnaeus's emendation were not immediately obvious. I would like to suggest that the semantic valence of *luridus* was both deliberate and significant, as Linnaeus wanted to provide a color term that was not simply an "intermediate" shade between white and black, but one that was specifically able to suggest a color of ill health or disease. Bernier also pointed to the dual meaning of yellow skin when he stressed that the women of India embodied a beautiful *jaune* rather than a jaundiced one. Linnaeus, for whatever reason, decided to classify his *Homo asiaticus* as the latter kind of yellow, and in doing so he was undoubtedly thinking within his two real areas of expertise: medicine and botany.

In the last chapter I briefly mentioned that both Chinese and Japanese whiteness was sometimes characterized as sallow, sickly, or deathly pale. But in medical discourse yellow skin had always been most closely associated with jaundice, as the English word itself, derived from the French for yellow, suggests. Yellow was an important color in humoral theory, too, a tradition that began with Hippocrates and Galen and was subsequently disseminated throughout Europe via the tenth-century Islamic physician and philosopher Avicenna. As the color of yellow bile, yellow was generally associated with a fiery and choleric temperament. Yellow jaundice, also considered a disease of the bile or gall bladder, was typically described in eighteenth-century medical texts as an obstruction that allowed the bile to return to the blood and eventually to the skin, producing a yellowish color. Symptoms of the disease included lassitude, sloth, and laziness.[19]

We might be tempted to say that these adjectives bear a curious resemblance to the incipient sinophobic stereotype of the "lazy Chinese," but some notion of lethargy or slothfulness was a supposed feature of many other (if not all) non-European peoples, and a lack of energy or vigor is a symptom of many other (if not all) diseases of the body. There is, however, an interesting connection to India in the so-called Paris text of Mandeville's *Travels* (ca. 1400), which reported that those who dwelled near the river Inde were of a green and yellow (*jaune*) color, "as if they were afflicted with jaundice" (*jaunisse*). In an essay on race from 1777, to which we will return in a moment,

Kant also suggested that while it might seem that the Indians' *olivengelb* color stemmed from a general condition among the people for stopped-up gall bladders and swollen livers, in fact their "inborn" color was equally jaundiced (*gelbsüchtig*) in appearance.[20]

While we have no solid evidence that Linnaeus had in mind yellow jaundice or any other specific medical condition when the color of Asians was altered from *fuscus* to *luridus* (although *luridus* did appear in classical Latin as a way of describing a jaundiced condition), there was one further connection to botany that, so far as I know, has gone completely unnoticed. This appeared at the bottom of the "Observations on the Vegetable Kingdom" in the first edition of the *Systema naturae* (and repeated in Linnaeus's *Fundamenta botanica* of 1736), where as the last of the "Vires vegetabilium" (the power or force or potency of plants), under the heading of color, the text noted that "*Red* always indicates acid, and a *pale yellow* [*luridus*] and *sad* aspect of the whole plant renders plants suspect." At first glance the "sad" and "suspect" nature of this color might lead us to believe that it referred to the condition of chlorosis (lack of chlorophyll, yellowing of leaves) that afflicts plants when they do not have the right soil minerals or when various diseases afflict them. In his *Philosophia botanica*, however, first published in 1751, Linnaeus identified an entire order of plants that he named *Luridae* and that were also characterized as "suspect": a group that included nightshade, hawthorn, foxglove, and the tobacco plant. In other words, "suspect" referred not only to diseased plants but to ones that were narcotic, poisonous, or otherwise toxic. In his *Clavis medicinae* of 1766 such a category was also listed under the heading of *Virosa* (i.e., stinking, fetid, rank things).[21]

GOOD VERSUS BAD YELLOW

And yet in the *Philosophia botanica* Linnaeus also implied that *fuscus* and *luridus* were actually two variants of the same thing. In a brief discussion of color terms used in plant names, Linnaeus actually defined *luridus*, which was included under the category *niger* and could refer to a variety of yellow, brown, purple, or gray colors, as *fuscus*. But this still does not answer the question why *luridus* should have been chosen to characterize *Asia*. Is it that Asia was being considered "suspect" or dangerous, as the fabled continent of exotic seductions and enervating luxuries, spices and silks and tea?[22] Was Linnaeus trying to suggest that the color of Asians was sickly, withering, jaundiced, or "sad"? Did he mean something more neutral like "dark" or "brown," as in the earlier term *fuscus*? Or was it that Asians (Chinese and Japanese, at least) were being thought of as luridly pallid, which would distance them from the true *albus* of *Homo europaeus* but at the same time maintain something of their formerly "white" color according to a long line of Western travelers and missionaries?

I have mentioned before that yellow, like most colors, could carry both positive and negative connotations, depending on context and the language in which they were expressed. Very roughly speaking, a "good" yellow seemed to be associated with light, the sun, and gold, while a "bad" yellow has traditionally carried suggestions of treachery, jealousy, and falseness. In medieval Christian art yellow was the color used for Judas and the Jews; in modern contexts yellow is still a color of warning, as in traffic lights or school buses; in modern English a yellow person is cowardly.[23] A convenient epitome of many of these associations appeared in J. W. von Goethe's book on color theory first published in 1810 as *Zur Farbenlehre*. Yellow, he wrote, the color closest to light, "embodies a serene, happy, and softly exciting character." But when imparted to "impure and coarse" surfaces the "beautiful impression of fire and gold turns to one of foulness; and the color of honor and joy is reversed to discomfort and disgust." The yellow garments of bankrupts, Jews, and cuckolds were all given as examples of this "dirty yellow."[24]

While none of this said anything about the possibility or the reality of yellow *skin*, Linnaeus's tenth edition also provided a new list of attributes for each human group, and the fuller picture we are given of "Asian man" might be imagined as coming closer to Goethe's notion of a "dirty yellow" (figure 5):

Asiaticus: luridus, melancholicus, rigidus.
Pilis nigricantibus. *Oculis* fuscis.
Severus, fastuosus, avarus.

MAMMALIA PRIMATES. Homo.

I. PRIMATES.

Dentes Primores ſuperiores IV, paralleli.
Mammæ Pectorales II.

1. HOMO noſce Te ipſum. (*)

Sapiens. 1. H. diurnus; *varians cultura, loco.*

Ferus. tetrapus, mutus, hirſutus.
Juvenis Urſinus lithuanus. 1661.
Juvenis Lupinus heſſenſis. 1344.
Juvenis Ovinus hibernus. Tulp. obſ. IV: 9.
Juvenis Hannoveranus.
Pueri 2 Pyrenaici. 1719.
Johannes Leodicenſis.

Americanus. α. rufus, cholericus, rectus.
Pilis nigris, rectis, craſſis; *Naribus* patulis; *Facie* ephelitica, *Mento* ſubimberbi.
Pertinax, hilaris, liber.
Pingit ſe lineis dædaleis rubris.
Regitur Conſvetudine.

β al-

(*) Nosse Se Ipsum gradus eſt primus ſapientiæ, dictumque *Solonis*, quondam ſcriptum litteris aureis ſupra *Dianæ* Templum. *Muſ.* ADOLPH. FRID. *Præfat.*

Phyſiologice: Te contextum Nervis, intertextum Fibris, Machina tenella, ſed adoleſcente in perfectiſſimam, facultatibus inſtructam fere omnibus pluribusque, quam reliqua cuncta. *Nudum in nuda humo, natali die, abjecit natura ad vagitus ſtatim & ploratum, manibus pedibusque deviciendum Animal cæteris imperaturum; cui ſcire nihil ſine doctrina: non fari, non ingredi, non veſci, non aliud natura ſponte. Plin. Vides itaque qualem vitam nobis rerum natura promiſit, quæ primum naſcentium omen fletum eſſe voluit. Seneca.*

Diætetice: Te ſanitate & tranquillitate, ſi noveris, felicem: *Moderatis* conſervandum, *Nimiis* deſtruendum, *Variatis* afficiendum, *Inſvetis* frangendum, *Conſvetis* indurandum; polyphagum Culina inſtructiſſima, per errores gratiſſima, igne vinoque horrenda. *Parvo fames conſtat, magno faſtidium. Seneca.*

Pathologice: Te tumidam usque dum crepueris bullam, pilioque pendulam in puncto fugientis temporis. *Nihil enim homine imbecillius terra alit. Homer. Nulli vita fragilior; nulli tot Morbi, tot Curæ, tot Pericula. Breve univerſum utique ævi tempus: Pars æqua morti ſimilis exigitur; nec reputantur Infantiæ anni, qui ſenſu carent; nec Senectæ in pœnam vivaces: hebeſcunt Senſus, torpent Membra, præmoriuntur Viſus, Auditus, Inceſſus, Dentes, Ciborum inſtrumenta. Plin. Sic magna pars mortis jam præteriit, quidquid ætatis retro eſt Mors tenet. Totum denique hunc, quem vides populum, quousque cogitas eſſe; cito natura revocavit & condet; Mors omnes æque vocat; iratis Diis propitiisque moriendum. Senec. II: 59.*

MAMMALIA PRIMATES. Homo.

Europæus. β. albus, ſanguineus, toroſus.
Pilis flaveſcentibus prolixis. Oculis cæruleis.
Levis, acutiſſimus, inventor.
Tegitur Veſtimentis arctis.
Regitur Ritibus.

Aſiaticus. γ. luridus, melancholicus, rigidus.
Pilis nigricantibus. *Oculis* fuſcis.
Severus, faſtuoſus, avarus.
Tegitur Indumentis laxis.
Regitur Opinionibus.

B 3 δ. ni-

Naturaliter: Te audacis naturæ miraculum, Animalium Principem, cujus cauſſa cuncta genuit natura, eſſe animal flens, ridens, melodum, loquens, docile, judicans, admirans, ſapientiſſimum; ſed fragile, nudum, ſuapte natura inerme, ad omnem fortunæ contumeliam projectum, alienæ opis indigens, anxiæ mentis ſollicitæque tutelæ, precarii ſpiritus, pertinacis ſpei, querulæ vitæ, tardæ ſapientiæ. Temporis annihilati emortui contemtorem, præſentis vividi dilapidatorem, futuri dubii æſtimatorem in vita volucri, pretioſiſſima, irrevocabili; ſic optima quæque dies miſeris mortalibus ævi prima fugit; alium ad quotidianum opus laborioſa Egeſtas vocat; alium Luxus incarcerat, altilemque ſuffocat; alium Ambitio numquam quieta ſollicitat; alius Divitiis, quas optaverat, metuit, & voto laborat ſuo; alium Solitudo, alium ſemper veſtibulum obſidens turba. Hic ſe habere dolet Liberos, hic ſe perdidiſſe; Lacrymæ nobis deerunt antequam cauſſæ dolendi. Sed quousque nos malos noſtra mala rapuere: circumferre pericula, ruere in ignotos, iratos ſine injuria, ferarum more occidere quem non oderimus, ſecundos optare ventos, quorum felicitas eſt ad bella perferre; parum videlicet ad mortes noſtras terra late patet. Senec. *Cætera animantia congregari contra diſſimilia, at Homini plurima ex homine mala. Plin.*

Politice: Te recti loco tenere errorem publicum factum, qui te vix enixum conſvetudinis larva induit, nutrit, educat, alit, regit, ſecundum quam honeſtus, fortis, ſapiens, moratus, pius exiſtimaris; Cum gubernatus Opinione vivas ad Conſvetudinem, nec ad Rationem. Te, inter perituros conſtitutum, cum nulli contigit impune naſci, beneficii loco petere, ut ultimus cervicem præbeas, dum interim, dulci fortuna ebrius, colligendo in craſtinum Lapillos fortunæ certas, verbis in furorem ageris, aliter theoretice pios odio perſequeris, quantosque tumultus excitas, non ut ſervias, ſed cui; nugis irrevocabile tempus conſumis; immortalia æternaque animo volutas, de feris nepotibus pronepotibusque diſponis, novas ſpes oblitus conditionis Tuæ ponis, cum interea longa conantem mors opprimit, at in agone oculos aperiens (ἄνθρωπος) ſomnium primo obſervas; *Sic vivimus ut immortales & morimur ut mortales. Senec.*

Moraliter: Te ſub rudi larva Ineptum, Laſcivum, Imitatorem, Ambitioſum, Prodigum, Solicitum, Aſtutum, Auſterum, Invidum, Avarum, *transformari* in Attentum, Caſtum, Conſideratum, Modeſtum, Sobrium, Tranquillum, Sincerum, Mitem, Beneficum, ſuis Contentum. *Uni animantium*

Figure 5: Three of the four varieties of *Homo sapiens*, from Carl Linnaeus, *Systema naturae*, 10th ed. (1758–59). Each variety is now distinguished by a list of descriptive adjectives in addition to color. "Asian man" has become *luridus*. The description of Africans appears on the next page, not shown here. Princeton University Library.

Tegitur Indumentis laxis.
Regitur Opinionibus.

[*Asian*: pale yellow, melancholy, rigid.
Hair black. *Eyes* dark.
Severe, haughty, avaricious.
Covered in loose garments.
Governed by opinions.]

It might be going too far to suggest that Asians' supposedly "melancholy" quality was reflected in their "sad" and "lurid" complexions, or that their "severe, haughty, and avaricious" nature was meant to suggest a "suspect" disposition. Note, too, that their once *fuscus* skin color was now said to apply to the eyes instead, which might also imply a certain duplicity or inscrutability, both of which became common stereotypes. In sum, we now have a vivid composite of a pale, black-haired, melancholy, haughty, enrobed, tradition-bound "Asian" drawn from a variety of racialized clichés.

We would like to know more about Linnaeus's sources. He prefaced his tenth edition with a list of some of the more notable collections of specimens he had at his disposal, as well as a list of students who had been sent to different parts of the world for plant samples. Finally there were unenumerated travel accounts and "Trivialia & Generica." It is well known that aside from an early trip to Lapland Linnaeus himself almost never traveled, but for many years he dispatched a constant stream of pupils abroad, and three on the list he provided were sent to East Asia via the Swedish East India Company: Christopher Ternstroem, who had died on his way to China in 1745, and Olof Torén and Pehr Osbeck, both of whom had arrived in 1750. Torén and Osbeck each published brief observations on the appearance of East Asian people, but they did not agree very well with the *luridus* of the *Systema naturae*. Osbeck's journal referred to the Chinese as *helt hwita* (completely white), although this was modified by the fact that southerners were darker: *mycket bruna* (very brown). Torén's diary, on the other hand, which was appended to Osbeck's account and composed in the form of letters to Linnaeus, said that Chinese men were *gulblek* (yellowish), while upper-class women were *blonda* and those of the lower classes were *solbrända* (sunburned).[25]

It is possible that Torén's description of yellow Chinese people influenced Linnaeus's choice of the term *luridus*, yet *gulblek* and *luridus* are not exactly the same color, and, moreover, in Linnaeus this was a complexion that was supposed to apply to the entire continent, not just China. Besides, Torén called the Javanese, who should also occupy the same racial category, *mörk-bruna* (dark brown).[26]

In any event Linnaeus's term *luridus* seemed to have a very short life. For while in the twelfth edition of 1766–68, the last one he produced before his death in 1778, *luridus* was retained, when this edition was translated into German in 1773–75 the color was changed to *braun* as if to return to Linnaeus's original choice of *fuscus*. This translation also featured new "explanations" including a passage identifying Chinese as "white with rosy cheeks," Japanese as "more yellowish" (*mehr gelblicht*), and Indians as "olive." Finally, a few pages later we are told that the Moors were black, the Asians brown, the Americans red, and the Europeans white or yellowish (*gelblicht*)![27]

Certainly these revisions were a bit quirky, and while in a certain sense this edition did return to *fuscus* it seems to me that the real problem was not color at all but a question of what *Asia* was, since it was clearly no longer seen as a serviceable category. This was even more apparent in the thirteenth (Latin) edition published in 1788–93 and edited by Johann Friedrich Gmelin, which was accompanied by a lengthy footnote suggesting that perhaps it would be better if human beings were distinguished in an entirely different way. What followed was a new arrangement in which, crucially, man was divided by color rather than by continent, and, equally crucially, the inhabitants of Asia (if we include Australia as well) were spread across three different racial divisions.

The "white" species, Gmelin suggested, included western Asia as well as Europe, in addition to northern Africa, Greenland, and the Eskimos. The second race, now called *badius* instead of *luridus* (i.e., reddish brown, bay colored), encompassed "the inhabitants of the rest of Asia" and obviously included what we now call East Asia. While it might seem as if Gmelin's choice of *badius* had backed away from Linnaeus's insistence on *luridus*, Gmelin also added that *badius* was *ex flavo fusci* (from dark yellow). The rest of Africa was black and the rest of the Americans were *cupreus* or copper-colored, and there was also a brand-new race that had taken over Asia's erstwhile designation *fuscus*: the inhabitants of the South Pacific and the Australian islands, only recently discovered, in addition to "many inhabitants of the Indies."[28]

FROM FOUR TO FIVE RACES

Gmelin did not come up with this new arrangement on his own, having been influenced by the most authoritative taxonomer of the post-Linnaean period, German physician and anatomist Johann Friedrich Blumenbach, who mentioned that Gmelin had simply adopted his own racial schema.[29] In fact there were some important differences between the two, but before we turn to Blumenbach we must first examine the work of Linnaeus's great rival in France: George-Louis Leclerc, comte de Buffon, author of the massive *Histoire naturelle, générale et particulière*, published in forty-four volumes between 1749 and 1804. In some sense Buffon's magisterial classifications of the entire natural world could not have been more different from Linnaeus's equally magisterial assertions, although both authors shared the same conviction that despite the seemingly infinite variety of human shapes, languages, and cultures, all mankind was composed of a single genus: *Homo*. In the world of eighteenth-century natural science this could by no means be taken for granted, as polygenic arguments were just as frequently being used as a means of (among other things) justifying the institutionalized enslavement of Africans.

Linnaeus and Buffon, however, conceived of the idea of human variety differently, since Buffon, to put the matter as simply as possible, considered Linnaeus's categories to be overly rigid and too permanent. This is clearly indicated by comparing Linnaeus's telegraphic geographical labels, color terms, and descriptive adjectives to Buffon's 150-page "Variétés dans l'espèce humaine," first published in the third volume of his *Histoire naturelle* in 1749. Indeed the key word here was "variety," as Buffon presented the reader with a positively dizzying array of overlapping nations, peoples, and cultures, with an equally vast variety of terms to express the color of their skin. He offered no table of the races but instead a summary of travelers reports (most of which were scrupulously cited) and personal communications, with only occasional suggestions about how different peoples might be divided and distanced from European whiteness, the true "primitive" color.[30]

The Chinese, he said, might or might not belong to the Tartar race, which was *basané & olivâtre*. While they might resemble the Tartars their manners and customs were entirely different. Some writers said they were a mixture of *basanez* in the south and *blancs* in the north, while others claimed they were "ash colored" (*couleur de cendre*), "naturally white," whiter than the Tartars, and "yellowish." The Japanese were yellower or browner, while the Ainu were less yellow than the Japanese. Vietnamese were more *basanez* but also *olivâtre*, and the people of Siam were "brown mixed with red," but some said "ash colored." Buffon made little attempt to arbitrate between these different opinions, and fictionalized accounts were placed side by side with eyewitness reports of every imaginable kind. Even George Psalmanazar, who would soon be universally recognized as an impostor, was cited as an authority on Formosa.[31]

The true cause for these different skin tones was also undecided, alternately described as an influence of climate, culture, living standard, disease, social rank, and exposure to the sun. One can easily understand, in a sense, how other eighteenth-century naturalists—most notably, Blumenbach—had become so dissatisfied with this apparent semantic chaos. But Linnaeus's rigid geographical boundaries were no better, and later in his career Buffon seemed to move closer to them himself. When discussing lions in a volume of the *Histoire naturelle* first published in 1761, he seemed to have accepted Linnaeus's general categorization of the human species as "white in Europe, black in Africa, yellow [*jaune*] in Asia, and red in America."[32]

Blumenbach, however, proceeded in a new way by appealing to physical characteristics instead of color, and to the shape of the skull in particular; his collection of human skulls was the largest of its day. In this sense he took a great step forward (or, rather, backward) in the development of scientific racism. His most important work on the subject, *De generis humani varietate nativa* (On the Natural Variety of Mankind), went through three editions and began as his medical dissertation in 1775. It appeared in expanded form in 1781, and then again in a completely rewritten version in 1795. At first only a small part was devoted to what we would now call race. Blumenbach agreed with Linnaeus that all human beings were composed of a single species and that there were important differences between mankind and the rest of the animal world. He also admitted that it was not easy to proscribe fixed topographical boundaries between the different varieties, and while he concurred that there were four main types, he was clearly unhappy with Linnaeus's separation of the world into continents. Instead, as Gmelin would echo a decade later, he defined the first category as those who live in Europe as well as in western Asia, Siberia, and part of North America. Their appearance and their color varied, but for the most part they "agree with us." The next group, who were found in the rest of Asia, were described as *fusci* or dark—much as in the early editions of Linnaeus. The third and fourth varieties included those who inhabited Africa as well as the rest of North America. Their color was not delineated either; as a matter of fact the only skin tone actually specified in Blumenbach's entire presentation was the "dark" people of eastern Asia.[33]

In his *Handbuch der Naturgeschichte* (Handbook of Natural History) of 1779, though, and then again in the second edition of *De generis humani varietate nativa* in 1781, Blumenbach expanded the number of human varieties from four to five, since as in Gmelin the people of the South Pacific were seen as a new race that must be accounted for. In 1779 they were called blackish brown (*schwarzbraun*), and in 1781 they were "intensely dark" (*intense fusci*), while East Asians, always the most difficult to colorize, were slightly lightened as "meist gelbbraun" (mostly yellow-brown) and then "subfusco plus minus ad olivaceum vergente" (darkish, more or less verging to olive). To make matters more complex, East Asia was further divided into northern and southern "stemmata," with the Japanese, Siberians, Manchus, and Tartars to the north; and the Chinese, Koreans, Siamese, and Vietnamese to the south. And as in so many other authors, East Asians represented a type that lay between the northern whites and the dark or black people of more southerly regions of the world.[34]

THE YELLOW MONGOLIAN RACE

By 1781, however, Blumenbach had also been influenced by a number of other contemporary taxonomies, including those of Buffon, Bernier, Leibniz, and Kant. Some readers may be surprised to learn that Kant was interested in the subject of race at all, but in fact his brief comments were highly influential and are worth examining, especially his remarks about a yellow Asia.[35] In "Von den verschiedenen Racen der Menschen" (On the Different Races of Man), published in 1775 and then in revised form two years later, the world was divided into four racial types: the "race of Whites"; the "Negro race";

the "Hun (Mongol or Kalmuck) race"; and the "Hindu or Hindustani race." Americans were included in the Hun variety (via migrations from northeast Asia), and the Malay peoples were placed among the black African race. The Hindustani peoples were olive yellow (*olivengelb*): the originary color of man or "the true Gypsy color" that explained the "more or less dark brown of other eastern peoples."[36]

As was so often the case, Kant characterized Indians rather than East Asians as yellow, and this was reemphasized in the revised version of 1777 with a new sentence attesting to the fact that "East Indians" were the result of a mixture of white peoples and the "yellow" Hindustani race. This revision was also concluded with a new summary of the races that oddly enough did not completely agree with the rest of the text. The white and black races remained more or less unchanged, but the Hun and Hindu types became "copper-red Americans" and "olive-yellow Indians," without any attempt to explain the discrepancy.[37]

The main difference, of course, was that the American race had replaced the "Mongol or Kalmuck" one, and this could perhaps be explained by the fact that for Kant color was the infallible marker of racial difference. But the Mongol variety was seen as a mixture of the yellow (Indian) and white (European) races, and this idea was repeated in Kant's essay on race from 1785, "Bestimmung des Begriffs einer Menschenrace" (Determination of the Concept of a Human Race), which noted once again that the color of Indians was "the true yellow" (*das wahre Gelb*). The Mongolian type, on the other hand, was determined by its form and not its color; it was only a subspecies instead of a true race. The evidence for this claim, such as it was, was an offhand remark in a popular travel text from 1776 that noted that children of mixed Russian and Mongolian heritage were born with "very handsome" (that is, not "half-breed") faces, which should not have been the case if the two groups really represented separate racial categories.[38]

But Blumenbach clearly disagreed with this view, and with each edition of his *De generis humani varietate nativa* he focused less on India (which he saw as a subgroup of the whites) and more on East Asia as the true center of the "dark" or "darkish olive" racial variety. By his third edition the review of previous taxonomers had also grown longer, including not just Buffon and Linnaeus and Kant but also Nicolle de la Croix, Thomas Pownall, Oliver Goldsmith, Johann Christian Polykarp Erxleben, and E.A.W. Zimmermann. Each of these authors gave a different account of between three and six human varieties, with some authors calling East Asians olive and others calling them white, and still others who repeated the idea that it was only the Indians who were truly yellow.[39] It is easy to see why by 1795 Blumenbach had become so frustrated by the question of color as well as the question of how "Asia" should be named and divided.

Another new term in Blumenbach's second edition, but which would become an absolutely central feature in the 1795 version, was degeneration. In taxonomic texts it had nearly always been implicit that the European race, whether or not it was specifically identified as white, represented the highest or most civilized or most perfect human form. White skin, in other words, was perceived as an *absence* of color, not "stained with pigment," as Blumenbach put it. He added that degeneration was most clearly demonstrated by the shape of the skull.[40] But in his third edition he also modified the five-race scheme into a spatially conceived hierarchy in which East Asians and Africans were seen as having degenerated the furthest from "primeval" white Europeans. Americans and South Pacific islanders, respectively, stood between these two extremes. East Asians, in other words, were no longer seen as an "intermediate" race between white and black; along with Africans they were now visualized as being at the furthest remove from white Europeans, with Americans and Malays in between (figure 6).[41] And the color terms themselves had also changed in a particularly complicated game of complexional musical chairs. While white Europeans and copper-colored Americans remained the same (although the Latin terms varied), the formerly *subfuscus* East Asians were now described as *gilvus* (pale yellow), the Malay peoples were called *badius* and tawny and *basané* rather than *fuscus*, and the *niger* Africans were called *fuscus* instead!

In the case of the newly canonized *gilvus* Mongolians, there was also an attempt to particularize their color by appealing to comparisons with food as well as to English terms presumably given by way of clarification (figure 7):

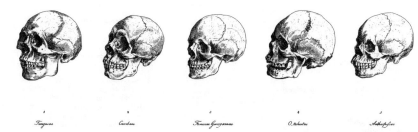

Figure 6: Five skulls, from Johann Friedrich Blumenbach, *De generis humani varietate nativa*, 3rd ed. (1795). The center skull, identified as a Georgian female, is judged as the most perfect and compared to the more "lopsided" skulls on either side. At one extreme, on the far left, is a "Mongolian" skull represented by a specimen from the *Tungusae*, a people of northeastern Asia. Princeton University Library.

> gilvus s. buxeus (angl. *yellow, olive-tinge*), medius quasi inter tritici granorum, et malorum cydoniorum coctorum, aut corticis exsucci et aridi malorum citriorum colorem; Mongolicis gentibus familiaris.
>
> [yellow or the color of boxwood (English: *yellow, olive tinge*), halfway between grains of wheat and cooked quinces, or the color of sucked out and dried lemon peel; familiar to the Mongolian peoples.]

Although a somewhat bizarre mishmash of clichés drawn from previous descriptions of other "intermediate" nationalities (*buxeus* had been applied to the Jews; quince-color was a standard descriptor for Native Americans), the basic meaning was clear: these people were *yellow* and should be thought of as such despite variations in different European languages.[42] He had probably chosen *gilvus* because it was closest to the German *gelb*.

While it is undeniable that this was our first unequivocal source for the idea of a yellow East Asia (a color, he said, that also included the equally fuzzy constellation of terms surrounding olive), it is absolutely vital to understand that this was not actually Blumenbach's most important or longest lasting contribution to the history of a yellow race. The real significance of Blumenbach's text was not that he decided upon yellow as opposed to any other color or that he had transferred an earlier notion of yellow Indians or yellow Semitic peoples onto the people of East Asia. Rather, it was far more important

SECTIO III. DE GENERIS HUM.

ve ulli genti adeo proprius sit, quin et in aliis, magnopere ceteroquin ab ea diversis, quandoque occurrat; in universum tamen omnes gentilitiae coloris varietates aptissime ad 5 sequentes classes principes referendae videntur.

1^mum^ scil. locum teneat albus color, qualis v. c. est plerorumque Europae populorum. Huic varietati genarum rubor fere proprius; in reliquis saltem rarius observabilis.

2. gilvus s. buxeus, (angl. *yellow, olive-tinge*.) medius quasi inter tritici granorum, et malorum cydoniorum coctorum, aut corticis exsucci et aridi malorum citriorum colorem; Mongolicis gentibus familiaris.

3. cuprinus (angl. *copper-colour* gall. *bronzé*.) s. obscure aurantius et

DEGENERATIONE IN SPECIE.

et quasi ferrugineus, cinnamomi aut coriario contuso cortici non absimilis; Americanis fere proprius.

4. badius (angl. *tawny*, gall. *basané*) s. medius inter ligni recentis mahagoni et caryophyllorum aromaticorum vel castaneorum colorem; Malaico stemmati et australis archipelagi hominibus communis.

5. denique fuscus, s. fuligineus, (angl. *tawny-black*) ad piceam usque nigredinem (angl. *jet-black*) in Aethiopicis quibusdam gentibus spectandam; quae vero fuliginosa nigredo neutiquam Aethiopibus propria est verum et in aliis maxime diversis et ab invicem remotissimis varietatibus generis humani, principali cutis colori admixta reperitur; ut in Brasilianis, Californiis z), Indis, et australasiis

z) De Brasilianis cf. v. c. G. Forster ad Wilson's

H 5

Figure 7: The colors of the five races, from Johann Friedrich Blumenbach, *De generis humani varietate nativa*, 3rd ed. (1795). The color of the second variety is given as "gilvus s. buxeus (angl. *yellow, olive-tinge*)" and compared to grains of wheat, cooked quinces, and dried lemon peel. "Common to the Mongolian peoples." Princeton University Library.

that he had simultaneously helped to invent a brand-new racial *category*—the "Mongolian"—which was every bit as fantasized and every bit as persistent as that much more notorious term he popularized at exactly the same moment: "Caucasian" (figure 8). These terms did not appear until 1795.[43] Yet it was not simply the case that in 1795 East Asians had finally become yellow; "Mongolians" had.

Once we understand that Blumenbach was thinking not so much in terms of color, the vagueness of which he repeatedly apologized for, but in terms of what would eventually be called physical anthropology, we can also better understand why Mongolians had shifted from their former position as an "intermediate" variety to a certain kind of racial opposite. The American type, which was seen as standing "between" Europeans and Mongolians on one side of the racial divide, was thought to be more physiologically akin to Caucasians

Sectio IV. generis hum.

rate ponderatis, universum, quout hactenus nobis innotuit, genus humanum aptissime ad ipsius naturae veritatem in *quinas* sequentes varietates principes dividi posse mihi videtur; nominibus

    A) *Caucasiae,*
    B) *Mongolicae,*
    C) *Aethiopicae,*
    D) *Americanae,*
et E) *Malaicae*

designandas et ab invicem distinguendas.

Caucasiam ob caussas infra enarrandas pro primigenia habendam primo loco posui

Haec utrinque in bina ab invicem remotissima et diversissima extrema abiit, hinc nempe in *Mongolicam,* illinc in *Aethiopicam.*
Medi-

species unica.

Medios vero inter istam primigeniam et hasce extremas varietates locos tenent reliquae binae:

Americana nempe inter Caucasiam et Mongolicam.

Malaica inter eandem istam Caucasiam et Aethiopicam.

§. 82.

Characteres et limites harum varietatum.

Sequentibus autem notis et descriptionibus quinae istae varietates in universum definiendae videntur. Quarum tamen recensui duplex monitum praemittere oportet, primo nempe ob multifariam characterum per gradus diversitatem non unum alterumve tantum sufficere, sed plurimis junctim sumtis opus esse; tum vero neque ipsum huncce characterum

Figure 8: The five races named, including "Caucasian" and "Mongolian," from Johann Friedrich Blumenbach, *De generis humani varietate nativa*, 3rd ed. (1795). The text explains that "Mongolian" and "Ethiopian" represent "the remotest extremes" from "Caucasian," with "American" and "Malay" falling somewhere in between. Princeton University Library.

than the square, flat, and broad Mongolian face that Blumenbach also described in great detail.

Thus while it might appear that he was attempting to pinpoint a yellow/olive color once and for all, he was actually just as interested in clearing the air surrounding the term "Mongolian" itself, which he said had been repeatedly confused with "Tartar" (as in Bernier, Buffon, Goldsmith, or Erxleben), a tradition that began with Marco Polo. But the Tartar peoples, he emphasized, properly belonged to the Caucasian variety. A few contemporary authors followed suit and agreed that the Tartars were fairer or more "Western" in appearance than other groups in China, but according to Blumenbach—and it is hard to overestimate the importance of this verdict during the next century and beyond—the Chinese and Japanese were merely subvarieties (*Unterarten*) of the Mongolian type.[44]

YELLOW MAN FROM THE EAST

Before we move to a fuller analysis of the fate of the notion of the "Mongolian," which will occupy us in the next three chapters, I would like to conclude with one last appearance of the term "yellow" in Blumenbach's work that has also not been noticed. It is particularly telling because it occurred in the context of Chinese people, although only in passing, and once again in a section emphasizing cranial form rather than skin color. Regarding his "second variety," not yet called Mongolian but simply the people "beyond the Ganges," Blumenbach commented:

> Their heads are usually oval, with flat faces, narrow eyes drawn up toward the external corners, small noses, and all those other things that are so well known from many pictures of the Chinese, as well as from china and pottery figures. Those Chinese whom Büttner saw in London were of this kind, and so was the learned botanist Whang at Tong (that is, "yellow man from the East") (*i.e. flavus ex oriente*) whom Lichtenberg recently met in that same city.[45]

Christoph Gottlieb Büttner was one of Blumenbach's teachers at Göttingen, but it is unclear exactly which Chinese people he had met in London, although there were several well-known visitors during this period. As for Whang at Tong (variously named in contemporary sources as Whang Atong, Wang-o-Tang, Whang-At-Ting, and Wang-y-Tong), somewhat more information is available. As a boy he was brought from Canton to England in the early 1770s to serve as a page to the Duchess of Dorset (or perhaps the Duke's mistress). He was educated at the exclusive Sevenoaks school adjacent to Knole House, the home of the earls of Dorset since the end of the sixteenth century, and by 1785 he had returned home to China to work as a trader. A perfect example of eighteenth-century chinoiserie (the Duke did not merely collect Chinese objects but counted a real-life native among his household), Whang was also the evocative subject of a beautiful painting by Joshua Reynolds that still hangs at Knole.[46]

The object of some notoriety, Whang made a number of appearances in eighteenth-century sources, as well as in a 1775 satire on

China by Lichtenberg himself. He also appeared in an essay by the famous Indologist Sir William Jones, who was intensely interested in Chinese as well. While in India in 1785, Jones wrote to Whang in Canton to ask for his help in translating selections from the Chinese classics and to encourage him to visit India with a group of other native speakers. Whang's polite reply, in which he said that such a translation would take many years to accomplish, was attached to Jones's essay published in 1799.[47]

But what is of such great interest in Blumenbach's passage is that he immediately translated the name Whang at Tong as "yellow man from the East" (and *flavus* not *gilvus*), a reading that can actually be traced to Lichtenberg himself, who wrote to Blumenbach in 1775 that the name translated as "der Gelbe aus Morgenland." Was this to suggest that his name actually *meant* "yellow man," or that it was simply a nickname or a reflection of some sort of contemporary stereotype about the Chinese as yellow people? If we are to understand it as the last of these possibilities it would be all but unique in sources from the period. The only other example I am aware of, briefly mentioned in chapter 1, was a slightly earlier Chinese visitor to England known as Tan Chitqua (or Tan Chetqua), a Cantonese artist who had arrived in 1769 and had achieved a great degree of success as a fashionable "face-maker." A 1770 description noted that his "complexion [was] different from any Eastern[er] I ever saw, with more yellow in it than the Negroes or Moors." Yet this same account was also accompanied by an extract from the *Gentleman's Magazine* that averred that "his face and hands [were] of a copperish colour." And in the second edition of William Chambers's *Treatise of Oriental Gardening* (1773), that famous proponent of Chinese architecture and gardens, Tan was introduced without any reference to his coloring.[48]

I suspect, rather, that "Whang at tong" was simply a mangled Romanization of some combination of the family name Huang, the character that is also used to signify the color yellow (as in Huangdi, The Yellow Emperor, a legendary figure also widely known in early modern Western sources), and perhaps Dong, the character for East and a common element in Chinese forenames. Whatever the case, I find this example particularly resonant because it seems to run counter to our expectations about "yellow Chinese" in a post-racial world. For we might have expected Blumenbach to seize upon the detail as some sort of "illustration" of the yellowness of his second racial variety. Not only did he not do so, but in his third edition, when *gilvus* Mongolians were finally canonized, the reference to Whang dropped out completely. Lichtenberg did not make much out of the "yellow man from the East" either. In the satire in which Whang appeared, which took the form of a conversation between a European narrator and a Chinese mandarin, with Whang serving as interpreter, the young man was not described at all. The face of the mandarin, however, was characterized as looking like it had been "carved out of meerschaum, and nearly the same color only somewhat greener."[49]

Perhaps, though, it would be better to say that in Blumenbach's third edition the reference to Whang was simply reincorporated into a much longer fifty-page section on what Blumenbach called "national faces" (*faciei gentilitiae*), which is to say the degree to which the shape of the face actually differed across human varieties. In the first edition there was a tentative discussion about the relative importance of climate as opposed to other factors such as the mutilation of the skull by artificial means, especially by "uncivilized" nations. The Chinese, he claimed, "are the least contented in the natural shape of their bodies, and therefore by many artificial means they distort and squeeze it, so that in many of their parts they differ from all other men." The custom of footbinding naturally comes to mind, as well as other theories that flat faces (and not only among the Chinese) could be caused by mothers carrying their children on their backs. At the same time Blumenbach also cited a description of three recently acquired Chinese skulls in Paris, which, the report claimed, could not be distinguished from the skulls of Europeans.[50]

But in his third edition Blumenbach appeared to speak with much greater confidence about the "racial face" of the Mongolian type (and not "Tartar," he once again emphasized), with an equally confident assertion that this racial face could be recognized even in abortions. While the question of climate still remained, the Mongolian "almost square" face was also now firmly paired with the Ethiopian as one of the two most "extreme" forms with respect to the "perfectly symmetrical" Caucasian head.[51] As always, skin color had become far less important than more easily quantifiable factors such as the shape of the face; the supposedly measurable "flatness" of human skulls, as well as variations in all other parts of the body, would become an obsession for physical anthropology throughout the nineteenth century.

Theorists after Blumenbach still sometimes puzzled over whether it was most appropriate to call East Asians yellow, olive, or some other hue, but a generalized idea of yellowness was clearly beginning to take hold, and the new notion of the "Mongolian" even more so. Eyewitness sources continued to disagree with regard to the Chinese, for example, who were still regularly being characterized as more or less "fair" depending on their location, their gender, and whether they spent the majority of their time outdoors.[52] But after Blumenbach and the other taxonomers it was no longer just a question of how to colorize an Asian race; the problem was how to colorize a "Mongolian" one.

Chapter 3
Nineteenth-Century Anthropology and the Measurement of "Mongolian" Skin Color

We concluded the last chapter by arguing that Blumenbach's 1795 resolution to call East Asians "yellow or olive" was the product of a long descriptive and taxonomic tradition, and that deciding upon *gilvus* to characterize the people of the region was a kind of reorienting (pardon the pun) of a constellation of color terms that had formerly been applied to very vague notions about the people of Asia as a whole. Blumenbach had attempted to correct earlier sources not only by settling upon yellow as opposed to some other color, but also by zeroing in on *East* Asia as its true source. Yet I have also tried to show that skin color was only one part of this story, since when Blumenbach created a five-part hierarchy with "Caucasian" in the center he simultaneously fashioned a brand new category called the "Mongolian," which he placed along with the "Ethiopian" at the furthest remove from his European ideal. I have insisted on the importance of this development because in the decades that followed, while scholars and travelers continued to disagree about how best to describe the color of the people of the East, their "Mongolian" nature was now almost universally taken for granted. A new *racial* category had been born, more defining and constrictive than skin color alone, and it would be most fully developed by the incipient science of anthropology, which will be our focus in this chapter. Early anthropology was overwhelmingly concerned with questions of race, and racial distinctions could supposedly be proven by collecting and comparing every imaginable variety of human measurement, mainly skulls at first but soon extending to all aspects of the human body. Even in Blumenbach the distinctiveness of East Asians was based on their anatomy just as much as their supposed skin tone.

In the first decades of the nineteenth century this notion would be taken over by the comparative anatomist and zoologist Georges Cuvier, probably the single most influential figure of the period. He simplified Blumenbach's schema by positing only three main types, white/Caucasian, yellow/Mongolian, and black/Ethiopian, but he also began to argue that the three races were permanent, that they had developed in isolation, and that they could be placed in an even clearer hierarchy with Europeans at the top and black Africans, who were closest to the apes, at the bottom.[1] This was part of a larger movement within natural science away from cultural and environmental explanations of human difference toward a growing confidence in biology and heredity. Race came to be seen not as the product of such temporary effects as climate or cultural practice but something much more permanent and biologically determined. Blumenbach had been obsessed with skull shapes as markers of different human groups, but the new sciences of the first half of the nineteenth century—physiology, phrenology, ethnology, and, ultimately, anthropology—also began to posit a correlation between cranial shape and the levels of intelligence supposedly housed within, with more "animalistic" forms coming to be seen as signs of inferior mental faculties.

This was just a short step away from arguing, and it would soon enough occur, that the races represented different *species* that did not even share a common origin. Naturally this was a religiously heterodox position, as it implicitly rejected the biblical explanation for the spread of the human family from a single pair. While many theorists of the period, and notably Cuvier himself, remained convinced that the human race still formed a single species, polygenic notions became increasingly attractive in the nineteenth century, and not just as a pretext for attempting to rationalize slavery and imperialism. The presumed superiority of the West was also at stake, and it has been repeatedly shown in modern race studies that while the debate between monogenesis and polygenesis was absolutely ubiquitous, it cannot be reduced to simplistic notions about the political stance of individual authors or whether or not they happened to be sympathetic to the abolitionist movement.[2]

To some degree it was surprising that color classifications should have been retained at all, but this was indicative of a trend within Western thinking about racialization that needed to distance everyone else from the whiteness (which is to say beauty, culture, intelligence, and level of civilization) that was reserved for Europeans only (and, later, Americans—not the indigenous peoples but their European colonizers). Whether or not the idea of polygenesis was specifically invoked, whiteness would soon be described as an *absence* of color, as in Lorenz Oken's seven-volume natural history of 1833–42, the climax of which was an explanation of the different "levels" of mankind, in which the "yellow Mongolians of Asia" were placed just below the top. Yet it was only among "transparent or white" Caucasians that color finally "disappeared."[3]

East Asians, we will remember, used to be white as well. But now the whole notion of skin tone had become inextricably linked to scientifically validated prejudices and normative claims about higher and lower forms of human culture. Mongolians could not be white, but what exactly were they? In some of the earliest anthropological studies the question of color had faded into the background, although yellowness was frequently retained as a term of convenience, as in Cuvier. But the problem of how best to define this yellowness had not gone away, even as the racial category of the Mongolian had become more and more codified. If the rest of the human body could be quantified, why not color as well? The subfield known as physical anthropology was particularly interested in this question, and some of its most influential proponents attempted to solve it first by color matching (that is, by comparing skin samples with color scales and noting down the closest match), and later by devices supposedly able to measure the actual color of human skin. We will see at the end of this chapter that the desire to find yellowness in the people of East Asia was so ingrained in the Western imagination that some anthropologists tried to prove that their skin really *was* yellow.

MONGOLIAN VERSUS TARTAR

But before we begin to analyze these developments we must first pause to consider the category of the Mongolian in greater detail. For we have not yet begun to answer the question of why this term should have been selected to represent the people of the Far East. When Blumenbach chose it at the end of the eighteenth century it was actually a relatively new umbrella term for the region. Previously, whenever the many peoples of northeastern Asia had been lumped together they had most often been called "Tartar," a tradition that dates back to the very first medieval travel narratives of Marco Polo, John of Plano Carpini, and William of Rubruck, all of whom referred to the Great Khan, leader of Cathay and founder of the Yuan or Mongol Dynasty, as a Tartar prince.[4] The term "Mongol" was certainly not unknown (a classic description appeared in Vincent of Beauvais's *Speculum maius* in the middle of the thirteenth century, drawn from Carpini's narrative, and then reprinted in such texts as Hakluyt's *Principal Navigations* in 1598), but Europeans had long used "Tartar" (or "Tatar") and "Tartary" to describe all the peoples north and west of the Great Wall, including the Mongols, the Manchus, the Kalmucks, the Buryats, the Tibetans, and many others. Even Carpini's narrative was titled "the history of the Mongols, by us called the Tartars." Their reputation as inhuman invaders was aided by a convenient pun on Tartarus, the Latin word for classical Hell.[5] Previous to the appearance of Juan González de Mendoza's landmark description of the Middle Kingdom in 1585, in fact, China itself was often called Tartary, as in the first edition of the earliest Western world atlas by Abraham Ortelius (1570), where there was no entry for China at all but for "Tartaria or the Kingdom of the Great Khan" instead.[6] Martino Martini's best-selling account of the Manchu conquest, which resulted in the foundation of the Qing dynasty in 1644, was called *De bello tartarico*, and Jean-Baptiste Du Halde's equally influential encyclopedic account of 1736 frequently mentioned China's "Tartar" government.

Later in the century, however, particularly at the instigation of Catherine the Great, who became empress of Russia in 1762 and who wanted the easternmost regions of her empire explored and brought under her control, a large body of new material was published in cooperation with the Imperial Academy of Sciences in St. Petersburg, including Johann Eberhard Fischer's *Sibirische Geschichte* (Siberian History) (1768); Peter Simon Pallas's *Sammlungen historischer Nachrichten über die mongolischen Völkerschaften* (Collection of Historical Information on the Mongolian Peoples) (1776); and Johann Gottlieb Georgi's *Beschreibung aller Nationen des russischen Reichs* (Description of All Nations of the Russian Empire) (1776–80). Each of these authors decried the inaccuracy of the term "Tartar" and how it masked a tremendous variety of peoples and cultures. But Pallas's work had perhaps the greatest effect, almost singlehandedly ushering in a new focus on "Mongolian" as an ethnic term for the entire region. We briefly mentioned in the last chapter how he had influenced Kant at an important moment in his thinking about the concept of race (and about East Asians in particular), but the effect of Pallas's encyclopedic analysis had a noticeable effect on Blumenbach as well.[7]

This effect was achieved mainly through the work of Christoph Meiners, professor of world history and Blumenbach's colleague at the University of Göttingen. His *Grundriß der Geschichte der Menschheit* (Outline of the History of Man), first published in 1785 and then in revised form in 1793, attempted to divide humanity into only two races, the "Caucasian" and the "Mongolian," both of which Blumenbach borrowed for his own five-race scheme. The Caucasian idea was an old one, centering around hypotheses about the location of Mount Ararat and the subsequent spread of Noah's progeny; it was a central thesis in E.A.W. Zimmermann's *Geographische Geschichte des Menschen* (Geographical History of Man) of 1778–83, also very influential in the German academic community. In Meiners's 1785 interpretation, however, the two races were clearly distinguishable in terms of both their appearance and their level of culture. A long list of essentialized differences followed, including not only complexion, body shape, hair, and stature, but also memory, understanding, imagination, diet, government, marriage customs, and codes of honor.[8]

"Tartar" peoples, significantly, were now placed in the "Caucasian" family instead (a distinction that would also appear in Blumenbach), but in Meiners's view the most important sign of human difference centered on "beauty" or "ugliness." Only the Caucasian race "deserves to be called beautiful, while the Mongolian is rightly called ugly." Both the Chinese and Japanese were said to be of Mongolian descent, and both nations were composed of dark-skinned "half-enlightened" people with malformed Mongolian body types.[9] In Meiners's second edition of 1793, though, the question of human beauty was given an even more fundamental position, since he replaced his original distinction between Caucasian and Mongolian with an assertion that it would be better to divide the races according to "white and beautiful" as opposed to "dark and ugly." Interestingly, moreover, Mongolian skin color was also silently changed from brown to yellow.[10]

In terms of a broader history of racialized thinking, Meiners's insistence on (un)attractiveness was undoubtedly a new low, as it is difficult to imagine a more blatant example of an attempt to fashion correspondences between whiteness and beauty, whiteness and culture, and whiteness and all forms of "civilization." It was hardly surprising that this view should have appealed to those who argued that the enslavement of Africans was somehow natural, or that he should have appealed to Nazi propagandists more than a century later, particularly as he claimed that within the white and beautiful race the most white and beautiful was the family of Germanic peoples. Yet what concerns us here is his more immediate effect on Blumenbach's "Mongolian." The two scholars, in fact, participated in a long debate about how best to divide the races, and Blumenbach rejected what he felt to be Meiners's overly impressionistic value judgments, which as in Buffon were based on a largely undifferentiated variety of travel reports, missionary letters, and other materials.[11] Certainly the shift to beauty and ugliness estranged Blumenbach even further, even if Meiners's alterations were incomplete, as Caucasian and Mongolian frequently appeared in his revised text as well.

But once both Meiners and Blumenbach had designated all East Asians as Mongolian, the character of the entire region began to acquire "Mongol" characteristics, especially the idea of barbaric hordes and merciless slaughter, centering on Attila the Hun, Genghis Khan, and Tamerlane, that familiar triad of invaders who were now regularly called "Mongols" as well. But the Mongolian was also considered a wandering and potentially dangerous human type that perhaps threatened to overrun the world once again, a fear of the populous East that would be invoked by such thinkers as T. R. Malthus, whose contemporaneous theory of overpopulation and limited food supply ignited the specter of a new "Northern Emigration," reminiscent of the invasions of Attila and Genghis Khan.[12]

Second, the Mongolian had also by this time become a common symbol of monstrosity and, in Meiners's conception, "ugliness." This, too, had a long tradition in the European imaginary. Mid-sixteenth-century encyclopedic sources called them *deformissimi*. In 1735 the Jesuit missionary Dominique Parrenin remarked how even southern Chinese who had moved to Tartary had became "true Mongols," with sunken heads, crooked legs, and a repulsive air of coarseness and filth.[13] Another "Mongolian" people, the Kalmucks, had been regularly singled out for their repulsiveness, as in Jean-Baptiste Tavernier's popular travel narrative first published in French in 1676, which twice paused to remark that they were "the ugliest and most deformed people under the sun" as well as "the most hideous and brutal," a verdict given even wider circulation when it was repeated by Buffon.[14] Petrus Camper, too, famous for the development of a "facial angle" to hierarchize the shape of human crania according to race (although this was not his original intention), had singled out a Kalmuck skull, which he considered singularly unattractive, to represent Asian man.[15]

Finally, in the 1780s universal historians such as Johann Gottfried Herder began to theorize that "Mongol" barbarism characterized all the peoples on "the Asiatic ridge of the earth," and that their "misshapen" and predatory nature was not simply climatological or cultural but hereditary as well. The Mongolian, that is, had by the time of Blumenbach's third edition become a key feature of the supposedly backward and unprogressive Far East in general, and this, too, was based on a relatively recent stereotype in European perceptions of East Asia—as Western sinophilia and its obsession with chinoiserie had gradually given way to an often equally virulent sinophobia in the second half of the eighteenth century. East Asian civilizations, although they had achieved greatness well before the Judeo-Christian West, were now regularly judged to be at a standstill, epitomized by Herder's notoriously negative picture of China as "an embalmed mummy, painted in hieroglyphics and wrapped in silk." And its "Mongol" characteristics were repeatedly blamed for this stasis.[16]

"MONGOLIAN" EAST ASIANS

Blumenbach's choice of "Mongolian" for the third and final version of his *De generis humani varietate nativa*, in short, was not merely a term of convenience like "Malay," which he admitted was more of a linguistic category than anything else.[17] Named after the Mongols of the thirteenth century, "Mongolian" brought with it notions of a nomadic, powerful, barbarous, and invading race, which of course the idea of their yellowness, as was perhaps the case with Linnaeus's *luridus*, could also help to reinforce. Never mind that the Chinese and Japanese, at least in terms of their cultural characteristics, had practically nothing "Mongolian" about them. And it is telling that while subsequent theorists continued to disagree on the most precise color terms for East Asians, or indeed on the degree to which Blumenbach's whole apparatus might be modified, enlarged, or even dismantled, they unanimously seemed to support the reality of a Mongolian racial stock.

In the subsequent thirty or so years, to give a few examples, the newly canonized Mongolian peoples were variously described as brownish red and yellowish, yellow and olive, brownish yellow, and "very pale yellow orange entering into pistachio green." Kantians continued to argue that it was the inhabitants of (modern) India who embodied the true yellow.[18] In 1824 Julien-Joseph Virey accepted the idea of a *race jaune* but preferred to categorize it among the white species instead (he said there were only two: white but never black and black but never white). Conrad Malte-Brun's universal geography chose to call Mongolians "the original race of the ancient continent," and Karl Rudolphi's textbook on physiology rejected Malay as a separate race, but both scholars agreed that East Asians were yellow.[19] In terms of the number of races, a useful summary of some alternatives appeared in Jean Baptiste Bory de Saint-Vincent's *L'homme* of 1827.

In 1806 André Duméril hypothesized that there were six racial types; Cuvier said there were three, and Virey only two. In 1825 Antoine Desmoulins suggested eleven, and Bory himself decided upon fifteen, including such curiosities as Japhetic Europeans, Hyperboreans, and Neptunians.[20] While there was certainly disagreement over just how many human varieties there were (one ambitious contemporary suggested thirty-seven), it is notable that "Mongolian" appeared in each of these authors, with the exception of Bory himself, who neglected Blumenbach entirely. Even early sinologists like Abel Rémusat, who certainly knew better, begrudgingly retained "Mongolian" (and, in fact, "yellow") as terms of expediency.[21]

However, there were many problems with terminology that needed to be worked out. Were these races, types, varieties, or species, and how many of each was the most appropriate number? A telling case in point is the article on race appearing G. S. Mellin's *Encyclopädisches Wörterbuch der kritischen Philosophie* (Encyclopedic Dictionary of Critical Philosophy) in 1802, which unhelpfully tried to distinguish between *Nachartung, Abartung, Ausartung, Race, Spielart, Varietät,* and *besonderer Schlag*, all of which had appeared in Kant.[22]

But what about "yellow Mongolians"? How did they fit into the notion of polygenesis and into new nineteenth-century ideas about human typology? The simplest answer to this question was that as in the color of their skin they were seen as a kind of "intermediate" race between the white and the black. For Cuvier this became a convenient way to distinguish them as less "civilized" than Caucasians but not as "barbaric" as Ethiopians, since the civilizations of the Far East had "always remained stationary." William Lawrence made a fuller statement in 1819, arguing that while China and Japan were "susceptible of civilization" they had remained "stationary for so many centuries": although they may well "exhibit the singular phenomenon of political and social institutions between two and three thousand years older than the Christian era," their subsequent stasis "marks an inferiority of nature and a limited capacity in comparison to that of the white races."[23]

Debates about race during the period immediately following Blumenbach were thus concerned not merely with anatomy but with much more dangerous notions about "inferior natures" and "limited capacities"—even if, to a large extent, such prejudices had long been implicit. Linnaeus had included sets of vague adjectives to characterize the personality traits of his four geographical varieties, but Blumenbach specifically downplayed this tendency by focusing on exterior form only. The great taxonomists of the eighteenth century had struggled with gradations in human skin color and cranial form, but nineteenth-century natural scientists began to suggest new and supposedly permanent hierarchies among the races themselves, an idea discernible even in staunchly monogenic thinkers like Lawrence, who noted that although all people form a single species they differ both physically and in terms of their "moral and intellectual qualities." Mongolians, like Africans, were barbarous but their character had also undergone some "softening." Whether or not human beings were composed of different species, it was only the white races who were said to be truly brave, free, compassionate, and benevolent.[24]

And now that Chinese and Japanese were Mongolian, both Lawrence and Cuvier paused to remind their readers that this was an inherently destructive racial type. Lawrence's outburst was typical of the era:

> When the Mongolian tribes of central Asia have been united under one leader, war and desolation have been the objects of the association. Unrelenting slaughter, without distinction of condition, age, or sex, and universal destruction have marked the progress of their conquests, unattended with any changes or institutions capable of benefiting the human race, unmingled with any acts of generosity, any kindness to the vanquished, or the slightest symptoms of regard to the rights and liberties of mankind. The progress of Attila, Genghis, and Tamerlane, like the deluge, the tornado, and the hurricane, involved every thing in one sweeping ruin.

While this tirade was directed at the Mongolian hordes of the distant past, modern China and Japan were firmly said to be part of the same family—"peopled by races of analogous physical and moral characters," with a "short stature, olive coloured skin, deviating into lighter yellow."[25]

The erstwhile whiteness of East Asians was sometimes reinvoked, as in *Researches into the Physical History of Man* (1813) by James Cowles Prichard, another staunch monogenist, where the complexion of Mongolians "varies from a tawny white, to a swarthy, or dusky yellow or copper colour." "The more civilized people have a larger stature," he added in a footnote, "a better form and a lighter complexion," but this may be due only to their relative hairlessness, and "were it not for this particular defect . . . we should probably find [them] nearer the Negro in complexion." Thirty years later he concluded that there were also frequent fair-skinned "deviations" from the Mongol type (who were by now also said to differ "but little in hue from the yellow soil of their steppes"), but instead of questioning the racial category itself he chose to argue, like Lawrence before him, that in such variations the Mongol race had "become softened and mitigated."[26]

Others tried to separate the whites from the rest of the world through different means, such as Carl Gustav Carus's suggestion that human beings could be divided into "day" and "night" peoples, with two "twilight" groups (one eastern, one western) in between. Mongolian yellow was suitably intermediate. Another proposal came from cultural historian Gustav Klemm, who said that the races could be distinguished as either "active" or "passive"—with Europeans, naturally, being the most active and therefore civilized. East Asians, like most other non-European cultures, were predictably passive, yet the case of China was improved by the fact that for thousands of years its people had mixed with the "active" races as well, which might also explain why, as he noted, their skin color was the same as Europeans. Still, they were not European, not fully "active," not quite white. A number of writers similarly stressed the almost-whiteness of the Mongolian races in general; both Charles Hamilton Smith and John Bachman called them olive, with Smith adding that they were "never entirely fair nor intensely swarthy"; and R. G. Latham said they were "yellowish-brown," "rarely a true white, rarely a jet black."[27]

Even the Bible could be said to support such hierarchies, depending on a peculiar combination of metaphorical and perversely literal readings of the story of Noah. The sons of Ham, who were fated to be "the servants of servants," had long been identified with Africans, even though the Bible said nothing about color. Yet it was also said of the sons of Japheth that they would "dwell in the tents of Shem." While Shem had usually been identified with Asia, as in the word "Semite," some nineteenth-century readers such as Bachman argued that "the widespread Mongolians" were in fact the progeny of Japheth, "many of whom," he tersely noted, "to this day are dwelling in tents." Whatever the case, Africans were always the losers, and Mongolians, if they were mentioned at all, were destined to remain somewhere in between.[28]

More scientific proof of the middling Mongolian was provided by Samuel George Morton, a professor of anatomy in Philadelphia whose collection of skulls now numbered into the hundreds. In the 1830s he not only compared the facial angles of human crania but also began to measure them from the inside—proving for many of his readers, that is, that Caucasians had the largest brains and therefore the highest degree of intelligence. Mongolian skulls were accorded second place, although not by much. Later in his career the nonwhite races moved slightly up and down, with Malay actually seeming to overtake Mongolian by 1849, but this hardly mattered as long as Caucasians remained firmly at the top.[29] Although Morton himself had been cautious about the question of species, his work was the direct inspiration for Josiah Nott and George Gliddon's 700-page *Types of Mankind* of 1854, which became the most widely read polygenic text of its day, especially in America. Nott and Gliddon were unabashed in their acceptance of separate racial origins, which, they repeatedly stressed, could be proven by the fact that the races were distinguished even in the most ancient Egyptian and Assyrian monuments.

Chinese culture, in fact, was said to demonstrate the permanence of the races most conspicuously of all, since Chinese faces, as Nott contended, appear exactly the same way now as they did more than three thousand years ago. Morton had apologized for the fact that he possessed so few skulls from the "Mongolian group" (only six were listed in 1849, as against more than three hundred of Native Americans, his real specialty), but Nott attempted to counter this deficiency with illustrations of purportedly ancient faces taken from popular sinological books such as Guillaume Pauthier's *Chine* of 1837—as well as the so-called four races of man shown in Seti's tomb, which we examined in the introduction. This seemed to be enough to establish the "Mongolian permanence of type."[30]

Even worse, at exactly the same time Arthur de Gobineau's *Essai sur l'inégalité des races humaines* (Essay on the Inequality of the Human Races) pressed a number of aspects of racialized scientific thinking during this period to their logical conclusions. For he argued not only that racial distinctions were permanent, but, as he announced so clearly in his title, that they were fundamentally unequal as well. So much for the entire Enlightenment conception of progress and perfectibility, as well as its accompanying assumptions that the rest of the world could be genuinely civilized (if not whitened) by the introduction of Christianity. His manifesto was nothing less than a universal history based on the idea that only the white races were civilized; as one of his chapter headings put it, "history exists only among the white nations." Like Meiners before him, Gobineau also appealed to Nazi theorists, since he not only posited a superior Germanic family but also an "Aryan" race that was the source of every true civilization in human history—including China's, which he said was brought by an Aryan colony from India. Borrowing from Cuvier, he conjectured only three "pure and primitive elements of humanity," the white, the black, and the yellow, and although his conception of a yellow type was sometimes unusual (he hypothesized that it originated in America and then spread to East Asia from the north), as with most other authors its "mediocrity" was seen as patently obvious.[31]

MEASURING SKIN COLOR

It was only a matter of time before nineteenth-century science began to wonder the degree to which racial difference—including skin color—might be measured in greater detail. In this sense Morton's work was only the beginning, since for the rest of the century and beyond the new field of anthropology became positively obsessed with the idea of anatomical quantification, and not simply a few assessments of the skull or skull capacity. People's entire bodies were evaluated with hundreds of tiny measurements of every imaginable sort. Yet it was also typical of Western race preoccupations that this obsession almost immediately spread to the question of color, as if it were now the explicit aim (although, of course, for many it wasn't) to prove that some people were "more white" than others.

The year 1859 is a convenient point at which to begin, not only because it marks the publication of Darwin's *The Origin of Species*, which far from squelching the polygenesis debate once and for all (although the book barely mentions the question of race), actually seemed to strengthen the idea of even more rigid and scientifically proven racial hierarchies. The question of evolution or transmutation had been hotly debated since the eighteenth century, and while in many ways Darwin seemed to solve these problems, he also gave credence to even more daring notions about natural selection, competition, and "survival of the fittest." The essential separateness of the races, for many readers, was now indisputable and supposedly supported by scientific inquiry.

For also in that year the French surgeon Paul Broca, best known for his pioneering researches on the human brain, founded the Société d'anthropologie in Paris, soon to be followed by the Anthropological Society of London in 1863.[32] His contributions to early anthropology were immense, and particularly what came to be known as physical anthropology. He was personally responsible for some two dozen measuring devices, including craniometers, goniometers, and micrometric compasses, all of which were standard tools by the time of his death in 1880.[33] Less well known today are his contributions to color measurement, even though Broca's *table chromatique* was regularly used well into the twentieth century, sometimes with minor variations. This table, looking much like a sheet of paint or makeup samples, featured twenty colored circles (for eyes) and thirty-four colored rectangles (for skin and hair) lithographed onto a piece of white paper or card stock. It was first published as part of the "Instructions générales pour les recherches et observations anthropologiques" (General Instructions for Anthropological Research and Observation), which appeared in the Société's *Mémoires* in 1865 as well as in monograph form that same year. A second revised edition appeared in 1879, where the text regarding color remained nearly identical, and the color table was now divided into three pages owing to the small pocket size of the volume (color plates 6 and 7).

That is to say, the book was designed as a practical guide for scientific (and armchair) travelers, who, it cannot be overstressed, no longer examined cadavers or skulls in laboratories but took measurements on the *living*. Tens of thousands of white European subjects were measured as well, it must be pointed out, but in 1874 the British Association for the Advancement of Science published its own revealingly titled version, which also included Broca's table: *Notes and Queries on Anthropology for the Use of Travellers and Residents in Uncivilized Lands*.[34] There was no question, in other words, that physical anthropology was intimately concerned with the measurement of racial difference, and that manuals such as these were geared toward evaluating those who were regularly referred to as "savages."

The main purpose of Broca's guide was to instruct observers to measure the head, face, torso, and limbs of their subjects, the results of which could then be recorded in a form that appeared at the end of the book in both "complete" and "abbreviated" versions. These forms were also offered for separate sale in bundles of one hundred, and the results could then be sent back to the Société d'anthropologie for future publication. Also available were brochures with "special instructions" for travelers to certain regions such as Senegal, Peru, and Sicily (and the second edition offered one for Japan). In addition to anatomical measurements, the form asked for a few general remarks about the weight of the subject, the pulse rate, the shape of the lips and nose, beard growth, teeth, as well as the color of the skin (both covered and exposed parts), hair, beard, and eyes. The observer was to record the numbers of the corresponding colors provided on the color table. The abbreviated version greatly reduced the number of measurements for each category but still requested color information.[35]

The text went into some detail about how to match the color of the eye, but far less space was devoted to questions of skin and hair. In fact Broca claimed that the determination of skin color was much easier than the other two categories, since the eye was complicated by many factors such as the structure of the cornea, the pupil, and the iris, the effects of light, the fact that eyes were often not composed of

Plate 1: Procession of Egyptians, tomb of Seti I, from Giovanni Battista Belzoni, *Narrative of the Operations and Recent Discoveries within the Pyramids* (1820). Princeton University Library.

Plate 2: Procession of "Jews," tomb of Seti I, from Giovanni Battista Belzoni, *Narrative of the Operations and Recent Discoveries within the Pyramids* (1820). Princeton University Library.

Plate 3: Figures from the tomb of Seti I, from C. R. Lepsius, *Denkmaeler aus Aegypten und Aethiopien* (1849). The top row shows three groups of "foreigners" in a procession illustrating *The Book of Gates*; the bottom row left features a group of Egyptians. The figures supposed to represent "Jews" or "Asiatics" are split between the two rows (bottom right and top left). Lepsius was careful to distinguish this group as "yellow-brown." Princeton University Library.

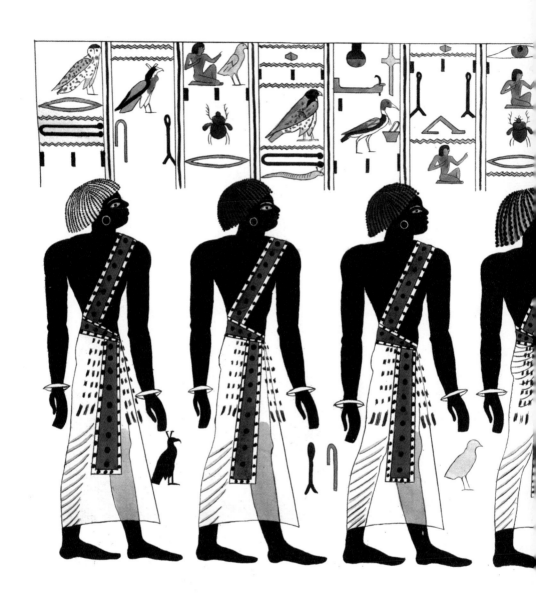

Plate 4: Procession of "Ethiopians" and "Persians," tomb of Seti I, from Giovanni Battista Belzoni, *Narrative of the Operations and Recent Discoveries within the Pyramids* (1820). Princeton University Library.

Plate 5: Figures from the tomb of Seti I, from Heinrich von Minutoli, *Reise zum Tempel des Jupiter Ammon* (1827). All the light-skinned figures are shown with the same skin tone. Princeton University Library.

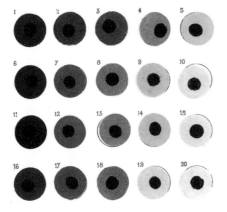

Plate 6: Chromatic table of eye colors, from Paul Broca, *Instructions générales pour les recherches et observations anthropologiques*, 2nd ed. (1879). Each row represents five shades of (from the top) brown, green, blue, and gray that become lighter as one moves to the right. Princeton University Library.

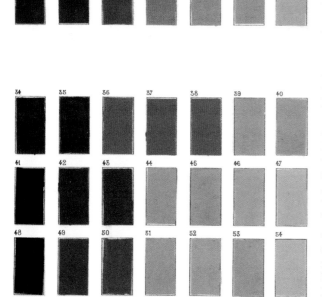

Plate 7: Table of hair and skin colors, from Paul Broca, *Instructions générales pour les recherches et observations anthropologiques*, 2nd ed. (1879). Each row begins with a shade of brown or black that becomes lighter as one moves to the right (although the colors have faded and are now difficult to read). Broca was vague about how these colors were chosen, except for number 48 in the lower left corner, which he said was pure black, and that the rows above it began with black mixed with red or yellow. The top row, however, seems to contain colors with no trace of black at all and thus begins one space to the right. Princeton University Library.

one uniform color, and so on. Hair was affected by light and not generally composed of a single color either. Yet the color of the skin, he claimed, was always flat and uniform, and the only caveat offered was that certain skins embodied a shiny and oily reflection that was not a serious difficulty.[36] No mention was made of developing controls for consistent color measurement, or even which part of the body should be utilized. Nor did anyone seem to be troubled by larger issues of color perception, in relation both to the physiology of the observer and to his or her language and culture.[37] Indeed the whole point of the color table was that it was supposed to provide international standards for colors by assigning them numerical values instead of conventionally vague and impressionistic descriptive terms such as "yellowish."

It turned out, however, that the many nuances of "yellow" and "red" skin were the most challenging. Once again it was the "in-between" shades that seemed to cause the greatest trouble, but this was not because yellow and red were really more difficult to judge. Rather, white and black were simply being taken for granted as self-evident entities. As one of Broca's most important successors put it, black and white were "primordial" and "incontestable," while red and yellow were "less evident, especially red." Little wonder that they were "less evident" since these were precisely the two colors that had been so arbitrary from the start, which is to say so much more than straightforward terms of objective description, even though "white" and "black" were equally fantasized notions. Centuries of observers had wrestled with the question of just what shade East Asians were, but by this time their Mongolian yellowness was a matter beyond question; as Broca had remarked in 1858, before the color table was even developed, the complexion of the Mongolian race was "more light among the Chinese proper than with the Samoyeds [i.e., a people in Siberia] . . . , but it is always yellow, and its variations are more limited than one can observe among peoples with white skin." Yet in the *Instructions* he admitted that when he first attempted to construct the table he was frustrated by the "multiplicity and proximity of the fundamental nuances relating to two colors only, the yellow and the red, as well as their exceedingly variable proportions when mixed together."[38]

In other words, yellow and red were not so simple after all, and it was clearly impossible to organize the skin and hair color samples into any sort of coherent order, as Broca claimed to have done with the twenty shades for eyes, which were arranged in five rows of four columns each, with each row representing five "degrees" of brown, green, blue, and gray. But how would one go about choosing basic skin and hair colors and then deciding upon their lighter and darker nuances? Perhaps one could start with black, brown, and blonde for hair (gray is already represented in the table for eyes), but what about skin? Obviously white, black, red, and yellow, the colors by which the races were almost universally known, were unworkable points of departure, since no one's skin was really any of these colors.

In fact this dilemma seemed to have caused considerable delay in the completion of the *Instructions*, which were announced as early as 1862, and Broca never fully explained how the resulting thirty-four colors had actually been chosen. He noted merely that the order in which they were laid out was not by chance, although it was not always a "regular" or "unbroken" sequence with respect to their "natural series." He did mention, however, that number 48 (in the lower left corner) was a pure black, and that while some observers claimed to have encountered people with completely black hair, the skin color of even the darkest Africans he had seen in Paris was always a shade mixed with red or yellow—colors that, he said, were represented at the beginning of the three rows above, numbers 27, 34, and 41.[39] In other words, each of the four rows at the foot of the table began with a shade of black (pure black, black mixed with yellow, black mixed with red, and black mixed with white or gray?) that became lighter as one moved to the right. This left only the row at the top, which seemed to contain shades with no black at all and were therefore positioned one space to the right. The effect of this arrangement was that when observers actually began to record their data it was only the white races whose skin tone matched the samples given in the top row. Although Broca certainly did not make a point of it, the table appeared to be laid out in such a way that, when it came to skin color, those with no trace of black in their skin were duly separated from the multitude of "colored" races that populated the four rows at the bottom.

Broca claimed to have overcome numerous material difficulties in the printing of the colors, including how to maintain consistency in each copy of the table, but the end result was a disturbingly confident tool for matching human skin, hair, and eye color in every part of the world.[40] As Edward Tylor noted in his 1881 *Anthropology*, one of the most widely used introductory texts of its day, "the traveler by using Broca's set of pattern colours, records the colour of any tribe he is observing with the accuracy of a mercer matching a piece of silk." While Tylor rejected the idea of matching eye color ("one has only to look closely into any eye to see the impossibility of recording its complex pattern of colours"), he enthusiastically embraced the idea of matching skin tones. Another early user was John Beddoe, notorious for hypothesizing an "index of nigrescence" for the races of Great Britain, the primary aim of which was to prove that the Irish were more "black" or "negroid" than the English. This index was based on hair and eye color, not skin. But in a different publication from 1890, in which he recorded some brief observations about the skin color of "tropical races" made during a trip around the world, he divided the skin of Indians, Chinese, native Australians, and Pacific Islanders into four "series" that were in perfect accord with the bottom four rows of Broca's table: red passing through reddish brown to black; orange or reddish yellow passing through brown to black, yellow passing through yellow brown and olive brown to black; and gray darkening to black.

The only term missing here was white, which was of course precisely the point. Beddoe admitted that a few of his Chinese subjects were the same color as the English, including some whose features were "almost European, with well-formed aquiline noses." He also admitted that he could not distinguish a group of English and Chinese workmen by the color of their bare legs alone. In an earlier period this might have led a Western observer to conclude that at least some East Asians were "as white as we are," but late-nineteenth-century anthropology was far more reluctant to give up its basic tenets of racial differentiation. If some Chinese were fair it might only indicate "the presence of an Aryan element," and as Broca's table had supposedly reaffirmed, "yellow preponderates."[41]

I have mentioned that the table was republished in England in 1874, and thirty years later an American version appeared in *Directions for Collecting Information and Specimens for Physical Anthropology*, a pamphlet published by the Smithsonian Institution and composed by Ales Hrdlicka, its first curator for physical anthropology and developer of a new subfield known as anthropometry. Interestingly, however, both the English and American versions soon scaled down the tables to fewer colors, a trend initiated by Paul Topinard in 1885. The second edition of the *Notes and Queries*, published in 1892 by the Royal Anthropological Institute, provided a simpler ten-color table instead, as also occurred in later editions of Hrdlicka's *Anthropometry*, first published in his *American Journal of Physical Anthropology* in 1919–20. But the sentiment was that Broca's table was still not practical or fully reliable, because the colors chosen were not useful for the particular group being tested, because color lithography was such an unstable media, or because it was difficult to isolate a particular color on the card as it was held up to the skin.[42]

Until these issues could be resolved and some sort of international agreement was in place, the guides recommended that an observer could resort to simpler and more impressionistic color judgments, which, as Hrdlicka noted, "in most cases are quite sufficient." He suggested a number of terms for each of the four "classes of color," which he identified as white, yellow, brown, and black. Note that yellow was presumed from the outset, even as red, that other bugbear shade, seemed to have been dropped as an acceptable descriptor. In later editions yellow and brown were actually combined and the number of terms was slightly reduced, but it is noteworthy that in both versions "light brown" appeared as a shade for the white as well as the yellow racial categories, since it was precisely the problem that Europeans and East Asians were often indistinguishably "light brown." In 1900 Joseph Deniker proposed a different set of ten "principal nuances" in which the white and yellow were more securely separated ("brownish white" as opposed to "yellowish white"), a distinction that was accompanied by a number of comparisons from other registers that the white category did not seem to require: the lightest yellows were compared to grains of wheat and described as *terreux* (i.e., muddy, pasty; in the English version: "a sickly hue"). The middle yellow (*jaune épais*, dull yellow, also called olive) was likened to "the color of new portmanteau leather," and dark yellow (*jaune brun*) was clarified as the color of dead leaves.[43]

Meanwhile other researchers tried to improve upon Broca's lithographed card by introducing new media. The most well known examples were Felix von Luschan's thirty-six colors of opaque glass, first developed around 1904 and reissued in the 1920s, and Gustav Fritsch's forty-eight tints in oil, produced around 1916. Others tried to replicate people's complexions by using watercolors and mixing them on site. One physician suggested leather samples as they more closely resembled the texture of the skin, and it was even claimed that toward the end of his life Broca had relied upon a table of more than one hundred colors produced by a glove factory. Some researchers developed more limited tonal ranges for particular peoples or regions, and others tried cutting out individual colors from tables like Broca's and mounting them onto separate strips.[44]

Something of a climax was reached in 1927 with Arthur Hintze's table of 358 colors mounted on fourteen cardboard plates, attached in the form of three large flexible fans that could be opened out to find the desired tint. Next to each color was a hole so that the fans could be laid directly on the surface of the skin. An initial match made with the "general" fan could then be refined with the "normal" (*sic*) and "red" fans designed to measure lighter and "middle" tones with greater precision. This was probably in response to numerous complaints that color tables were much easier to use with respect to darker skin, which was after all their primary purpose. Both Fritsch and Hrdlicka had remarked that tables lacked sufficient matches for "normal" complexions, and in the 1920s one researcher came up with a table exclusively for use on European skin rather than the "exotic races."[45]

THE COLOR TOP

But perhaps the oddest development regarding skin color during this period was an instrument known as the color top, which was actually a child's toy designed for the schoolroom to teach the fundamentals of color and color mixing. It was based on experiments in color and color perception conducted in the 1850s by James Clerk Maxwell, and color wheels had been around for centuries. Maxwell's great innovation was to attach colored disks to the wheel, with each disk having a slit cut from its circumference to the center that would allow the user to interlock and combine the colors in any desired proportion. When the wheel or top was spun the colors blended. "Maxwell disks" became a Victorian optical toy, and in the early 1890s, Milton Bradley, best known to later generations as the king of board games, decided that a simpler version would be beneficial for children's education. In 1895 the Bradley color top was patented and soon sold in Woolworth's at six cents each. It consisted of a wooden spindle attached to a thick cardboard disk, the outer rim of which was divided into units so that the percentage of each colored disk could be easily quantified. An assortment of paper disks were included: red, orange, yellow, green, blue, and violet, plus white and black (figure 9).[46]

Many anthropologists, surprisingly enough, adopted this toy as an eminently pragmatic (if not absolutely scientifically sound) research tool. This was especially true in the United States, partly because the top was more readily available there, but even more so because, as we shall see presently, it was seen as the most effective way of measuring blackness as opposed to any other color. The advantage of its portability was obvious, and because each of the colors could actually be quantified according to the gradations printed on the base of the top, it seemed far more accurate and copious than color scales such as Broca's or Luschan's. It first came to the attention of the scientific world as a way of studying the color of small animals and birds, and the major proponent of the method was biologist (and later eugenicist) Charles Davenport, who began his career teaching zoology at Harvard in the 1890s. An early example of the method was an 1898 experiment on the taxonomy of shrikes that argued that while the color top had detected traces of blue, red, and yellow in the birds, the most important measurement was the degree of black, or melanism, of the color areas concerned.[47] Soon thereafter, however, Davenport came under the influence of early eugenic theories and questions about Mendelian laws of heredity, and his attention gradually shifted

Figure 9: Color top, from Milton Bradley, *Elementary Color* (1895). Fig. 4 shows the gradations on the top and figs. 5–7 demonstrate how the disks are interlocked and adjusted to the desired proportion. The top was regularly used by anthropologists throughout the early twentieth century to measure skin color. Princeton University Library.

to the "breeding" of human beings. In 1904 he became founding director of the Station for Experimental Evolution, funded by the Carnegie Institution, where he began to amass an enormous archive of genealogical records for the purpose of eugenics research.

An early paper published in 1906, *Inheritance in Poultry*, began to take Mendelian questions of heredity seriously by mixing different breeds of chickens. But he also stated quite clearly in the introduction that the eventual aim of these experiments was to learn about cross-breeding between human groups. Davenport did not employ the color top here, although he frequently focused on many aspects of the birds' coloring. This was perhaps because he found little evidence of simple "blends" in his cross-bred samples, as was often the case with human beings. But by 1910 he was able to furnish the necessary data for a paper on human skin pigmentation, the results of which he divided into three sections: "ordinary" skin pigment in "typical Caucasians," the heredity of pigmentation in "crosses between whites and negroes," and the inheritance of albinism.[48]

Here he did make use of the color top, but revealingly enough only in the case of the black and white "crosses." For the "typical Caucasian" subjects it seemed to be enough to ask schoolchildren to fill out cards about their family history and to supplement this information with more detailed questionnaires and interviews. The color terms employed were basic and simple: blonde, fair, brunette, dark brown, and so on. Yet when it came to the question of "half-breed" or "mulatto" offspring, it was suddenly a problem that "the parentage of the children must be unquestioned" (the racial prejudices are obvious), and also that "the degree of pigmentation must be quantitatively expressed"—in other words, by means of a color top. In fact it seems that data for only four families had been obtained, but this dearth of information was then supplemented by a new and more specialized monograph of 1913, *The Heredity of Skin Color in Negro-White Crosses*, which purported to solve the problem by using data from Bermuda and Jamaica, since in much of the United States "all matings of blacks and whites are illegal." And the color top now seemed particularly useful since, as British statistician and early eugenicist Karl Pearson had argued in 1909, "60% of the children in some [West Indian] islands are born out of wedlock."[49]

Here is Davenport's description of the procedure involved:

> [The assistant] visited the homes of the colored people and obtained all the genealogical data that could be furnished. Then the sleeve was rolled up to above the elbow and a part of the skin that is usually covered from the sunlight was thus exposed. The arm was placed on the table by a good light and a Bradley color-top was spun close to the arm and the disks adjusted until they matched, when spun, the color of the skin. Various combinations of black (N), red (R), yellow (Y), and white (W) gave a close approximation to the skin color.

A recent commentary has referred to this form of measurement as an inadvertent self-parody, and indeed it is hard to imagine that a five-and-dime child's toy would become such an important research tool

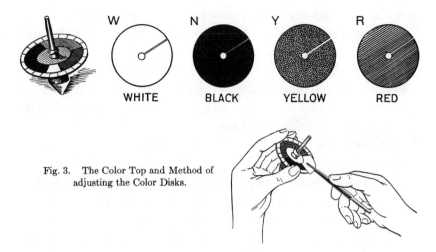

Figure 10: Color top, from Louis R. Sullivan, *Essentials of Anthropometry* (1928). The top as illustrated in a pocked-sized guidebook for the measurement of physical difference published by the American Museum of Natural History. Courtesy American Museum of Natural History. Princeton University Library.

(figure 10). Interestingly, a number of the top's proponents argued that one reason it was such a great improvement over older color charts is that there were no feelings of resentment when examinees saw that their complexion was being compared to the usually whiter skin of the researcher, especially in the case of Luschan's glass tablets, which for many anthropologists were preferable to Broca's scale since the colors were arranged in a clear progression from white to black.[50]

Later researchers suggested isolating the color of the skin sample with a white cloth or a piece of paper with a hole cut in the center, as well as another white material to cover the surface of the table on which the subject's arm rested. It was recommended that the disks should be adjusted with forceps since contact with the fingers tended to degrade the results. Field workers were also careful to bring along a sufficient number of tops as they easily became dirty or otherwise damaged by sunlight and humidity. As much as a 10% difference was detected between old and new tops. Some worried about what kind of light should be used and how to ensure that the top was spun at exactly the same speed every time (the color significantly darkened as the top slowed down), while others recommended that the color should be judged without too much deliberation in order to lessen the degree of subjectivity. Indeed many of these anthropologists seemed a bit wary of the procedure even as they continued to rely on it and to proclaim its results, mainly because, as they often noted, it was the best method available at the time.

But exactly what did it measure? It was soon recognized that the disks chosen (black, white, red, and yellow) were not necessarily the equivalents of the pigments present in human skin. The optical effect of mixing black and white to spectral colors makes them "broken," so all the color top really did was to produce a range or lighter or darker red-yellows, which is to say shades of brown and orange. And isn't all human skin one shade or another of those colors anyway? Yet it is nonetheless worth pondering why Davenport had selected these four disks in the first place. Basic color theory dictates that many different combinations could have been used to match the rather limited range that comprises the color of human skin, and there was also a confusion between the standard palette for mixing colored pigments (red, green, and blue) and for mixing colored light (magenta, yellow, and cyan). Davenport himself offered a rather half-baked and oddly retrospective explanation: that each color did represent an element in human pigmentation (black for melanin, yellow for "lipochrome" or carotene, red for blood, and white because it was "reflected from the opaque skin"). But he also claimed that the disks "indicated" that these were the pigments that compose skin color rather than the other way around.[51]

Other color top experimenters employed different disks for different purposes. A 1915 study on the relationship between yellow plumage and egg-laying fecundity in hens utilized the yellow and white disks only. Another paper analyzed the color of egg yolks produced with different kinds of feed and measured them with the white, orange, yellow, and green disks. A third experiment on guinea pigs chose the orange disk instead of the red.[52] It seems rather obvious that researchers began by selecting exactly what they initially "saw" and thus expected to measure, which is fair enough in the case of a yellow hen or the yolk of an egg. But what about human skin? There had always been considerable controversy over exactly what colors were present there, and even today many questions remain about both its physiological nature and its complex optical effects, sometimes referred to as "scattering."[53]

In Davenport's time though, red, yellow, and black pigments were generally assumed, and this probably accounts for his choice of those three disks. But the reasons for the white one were never really explained. I have just mentioned that his rationalization was that white was "reflected from the opaque skin," which, if I understand him correctly, was more of a claim about the effect of light than the presence of a color. And yet in a later paper he also argued that "the color of the skin proper, without pigment, may be regarded as white" (although white like a corpse)—a familiar prejudice about the whiteness of "normal" skin. As Topinard wrote in 1876, the three "fundamental elements" of the skin were red, yellow, and black, mixing in different degrees with "the white colorless base of the tissues." It would seem, in other words, as if the white disk represented Davenport's idea of "pigmentless" skin, and that high levels of yellow, red, and black only confirmed that the individual who was being measured was not Caucasian. When he examined a few South American albino subjects in 1924, they registered no black at all. And yet I would also argue that the simplest explanation for his color selection was that red, white, black, and yellow were exactly the colors that the "four races of man" were assumed to be, and we will see that this, too, would have surprising effects in the decades to come, especially on the descriptions of "yellow" and "red" people.[54]

The real focus of these initial experiments, however, was the blackness of the N disk (N for *niger* or *nigrum*), just as in the beginning the only people who were actually measured were black and white "hybrids"—as if the centuries-old classification of mulattos, mestizos, quadroons, and so on, might actually be quantifiable. It was no longer necessary to refer to dark skin as café au lait or chocolate-colored, both of which had been used extensively throughout the nineteenth century. As Luschan wryly noted in 1916, any skin would be the color of *Milchkaffee* depending on how it was prepared.[55] From a eugenics point of view the aim was to prove that skin pigmentation "segregated" and that it was passed on according to Mendelian laws of heredity. Davenport thought he had demonstrated that there were two biotypes of black African, one full-blooded and the other a hybrid, with the latter having much lower levels of black in the skin. When two mulatto parents had children their pigmentation "Mendelized" and both very white and very black children could result, particularly in the second and third generations.

Pure-blood "white" people, of course, did not need to be measured precisely because they were white. But the "crosses" could not be considered white even if their skin tone was as light or lighter than many Caucasians (even one drop of black blood made you black). Such people could merely "pass for whites."[56] It was for similar reasons that those who studied African American populations embraced the color top, since blacks in the United States were perceived as a racial crossing between "true" Africans and many generations of white and Native American interbreeding. In the 1920s the leader in this field was Melville Herskovits of Columbia, who used the top to support his argument that only about 20% of the black population in America was "pure Negro." The toy was likewise deemed useful for other mixed groups such as the so-called "white Indians" of Central and South America, and in 1914 Davenport journeyed to Australia on a government grant to measure aborigines and "black-white hybrids." As for East Asians, he lamented ten years later that "few quantitative studies have been made on the elements of skin color in Orientals," but this, too, would soon be rectified.[57]

Yet the top's ability to measure nonwhiteness may have been more than just a matter of racial prejudice. It may have been built into the nature of the very colors that were selected. In his primer on color theory first published in 1895, Milton Bradley himself noted that it was something of a mystery that the addition of only a little white changed the tone of a color drastically, while it required a great deal of black to produce a noticeable change (and that this is exactly the opposite of what occurs when mixing paint). It is perhaps for this reason that researchers noticed that the white disk was not really able to distinguish different Caucasian (and Mongolian) skin tones. The same had been said of Luschan's scale. Others complained that the yellow disk was just a mixture of red and green that merely produced a sensation of "yellow," or that the Milton Bradley company had changed the cardboard platform on which the disks rested from gray to brown, which threw everything considerably out of whack. Sometimes puzzling optical effects were recorded, as when a yellow object registered a higher percentage of red, or when a visibly lighter skin actually showed a larger amount of black than a darker one. Worst of all, there was as much as a twenty-point overlap between the amount of the N disk in darker Caucasians and lighter-colored blacks.[58]

It was then discovered in the early 1920s that the red disk wasn't even red at all (not spectral red, that is), since it was more of an oxblood shade that, when analyzed, turned out to contain 59% black.[59] The four colors on the top, ironically enough, were even blacker than Davenport had originally supposed, and all his early data needed to be recalculated if it was to be of any value at all. And yet the top continued to be used. In 1926 he became even bolder and published a popular essay in *Natural History* arguing that the skin color of all the races could be "describe[d] in exact quantitative terms" with the top. His justifications for the four colors also became more detailed: "the skin proper" was said to be a cream color that was perfectly gauged by the white and yellow disks; if the red disk was really oxblood so much the better, as the red pigment traceable in human skin happened to be just the same shade; and while melanin was actually brown it did indeed give the impression of pure black when it was concentrated.[60]

What made this essay different, however, was that Davenport also ventured into the much more uncertain waters regarding the "yellow" and "red" races, although he gave far more attention to the former than the latter. Since there were no earlier studies of East Asians, he measured the upper arm of a Japanese colleague, only to find that "the Y element is not exceptionally great" and that the readings of each of the four disks were almost identical to those of a group of Caucasian children he had just examined. Obviously this came as something of a surprise, particularly in light of the fact that "some other observations [he] has made suggest that Eastern Chinese show a large per cent of yellow in their skins." And without any evidence at all he concluded on the next page that if "the eastern Asiatics were formerly classified as the yellow race" it was "a terminology justified by the large size of the yellow factor."[61]

What clearer example could there be of the assumption that East Asians simply "were" yellow, even if it was not actually borne out by the results of investigation? A 1930 essay by Beatrice Blackwood of Oxford soon came to the rescue, since it had actually carried out color top measurements on a substantial number of each of the "four races." In addition to providing the most detailed review of the color top method ever published, Blackwood's essay also stands as an apotheosis of racial stereotypes "proven" by the four colors contained in the top—results that were achieved despite repeated apologies that "this technique does not give us an estimate of the actual amount of pigment present in the skin." A more literal-minded breakdown of who was black, white, red, and yellow could not be imagined, with numerical tables provided to show that blacks were blacker and whites whiter, that Chinese were second only to Caucasians in terms of their whiteness (although the gap was still large), that Native Americans were the reddest, and that the Chinese were more yellow than any other race, a fact that "conforms to traditional expectation." The color top demonstrated nothing less than that the names "Yellow Race" and "Red Indians" were both "expected" and "deserved."[62]

But at just this time the color top's star had also begun to fade, as new and more sophisticated devices for color measurement were being developed utilizing colored light (generally red, green, and violet) and with the aid of electricity. These devices went under a number of different names (the photometer, the spectrophotometer, the colorimeter, the photoelectric reflection meter), but in each case the principle was the same: they did not simply measure color in any simple sense but instead gauged the way that an object both reflected and absorbed light at various points in the visible color spectrum. In this sense spectrophotometry, as it was most often called, did not rely on visual judgment at all, thus avoiding many difficulties, but it is also revealing that just as in the case of the color top it was not long before someone decided that this new technology might also be used to classify human skin, the measurement of which, at least in the field of physical anthropology, was still perceived as "one of the most pressing and long-standing requirements."[63]

Spectrophotometry is still in use today in the field of dermatology, as it can be of great utility in the diagnosis of certain skin diseases, including jaundice. But in the 1920s and 1930s the device was seen to demonstrate with even greater scientific reliability the steady increase in pigmentation among the "dark races," from the lightest East Asian to the darkest African. And these researchers, too, regularly took the yellowness of East Asians for granted. A good example is a 1926 study that measured some two dozen subjects including a "yellow-skinned" Chinese man and a "yellow-brown" Malay woman. Even in the standard essay on skin pigments published in 1939, although the authors concluded that skin color was not due to "black," "white," "yellow," and "red" elements but just the level of melanin alone (and its derivative, melanoid), the term "the yellow races" still appeared.[64]

Whether they were "actually" yellow was no longer the focus, perhaps, but at the same time there could be no mistake that they were not white. The point of so much measurement was to show that the "other" races really did differ from the "normal" one, the group to which of course the anthropologists themselves generally belonged. Such differences were now firmly thought to be inheritable, and once again Davenport represented one of the more extreme cases of this sort of racialized thinking. In 1911 he published a 300-page monograph, *Heredity in Relation to Eugenics*, where everything from blue eyes to secretiveness to farming to walking quickly were seen as examples of "family traits."[65] The real thrust of these discoveries was to warn readers about the detrimental effects of the further influx of the "wrong sort" of foreigners, thus requiring a variety of immigration controls, legislation against undesirable marriages, and even mandatory sterilization. The threat of the "yellow" race, like all the other nonwhite ones, had become a problem.

But since modern "Mongolians" were after all just the children of a long line of wandering slaughterers from that part of the world, they were already being characterized as the most dangerous of all. The choice of Mongolian would prove even more fateful than yellow during the course of the nineteenth century, since the climax of

East Asian racialized thinking during this period, Kaiser Wilhelm II's coinage of the phrase "the yellow peril" to describe East Asian aggression in 1895, was a response to the region's "Mongolianness" as much as to its presumed color. This is the subject of chapter 5, but first we must turn to the question of the "Mongolian" in nineteenth-century medicine.

Chapter 4
East Asian Bodies in Nineteenth-Century Medicine
The Mongolian Eye, the Mongolian Spot, and "Mongolism"

In the last chapter we took up the idea of a yellow race in the field of nineteenth-century anthropology. This chapter will serve as a companion piece of sorts because it examines the fate of Blumenbach's Mongolian race in nineteenth-century Western medical discourse, even though medicine, unlike anthropology, did not typically make any claims about yellow skin apart from its manifestations in such diseases as jaundice or yellow fever. Western medicine did, however, attempt to strengthen the racialization of the region by employing the adjective "Mongolian" in a number of conditions that were supposedly linked to—or endemic of—the people of the Far East. The list includes the "Mongolian eye fold" (a fold of skin covering the canthus or inner corner of the eye, also called an epicanthus or epicanthic fold), "Mongolian spots" (congenital bluish marks appearing in infants on their lower back or buttocks), and "Mongolian idiocy" or "Mongolism" (now called Down syndrome or trisomy 21).

THE MONGOLIAN EYE

Each of these purportedly "Mongolian" conditions has since given way to more neutral names, although the racialized terms still crop up from time to time in different languages. By far the oldest of the three, at least in a general sense, is the "Mongolian eye fold," also known as the "Mongolian fold" and the "Mongolian eye." Although the terms themselves were not coined until the late nineteenth century, the idea that East Asians had "little eyes" can be found in some of the earliest medieval accounts such as Carpini's or Rubruck's, a detail that was then repeated in sixteenth-century reports even before travelers had actually reached China or Japan. By 1520 both Duarte Barbosa and Andrea Corsali had noted that the people of China had *occhi piccoli*. Both accounts were reported in G. B. Ramusio's groundbreaking compendium of voyage narratives published in 1554. The same detail also appeared in Sebastian Münster's best-selling *Cosmographia*, originally published in German in 1544, as well as in the first European book devoted exclusively to China by Gaspar da Cruz, published in 1569 and then widely disseminated in Juan González de Mendoza's popular description of 1585. Narratives of the first tentative encounters with Japanese people featured similar information, subsequently verified and repeated by early missionaries such as St. Francis Xavier.[1]

One might even argue that before the end of the eighteenth century, when the Mongolian race was first formulated, European commentators considered the most distinguishing physical characteristic of the people of the Far East to be the shape of their eyes, which along with their "flat" noses and black hair and relative beardlessness was far more often described than the color of their skin. It is also clear that Western authors struggled to find the right words to explain the "little" or "narrow" or "sunken" eyes that people of the region were said to possess. The pioneer missionary to China, Matteo Ricci, said that their eyes were not only small and black but "oval-shaped and sticking out," variations of which were attempted in different languages by Trigault, Martini, and Nieuhof, among others.[2] Eyewitness accounts of the group of Japanese boys who came to Rome in 1585 claimed that they had "protruding" or "acute" eyes, and a Chinese visitor to London in 1687 was characterized as a "narrow-eyed" "blinking fellow." Other commentators related that the faces of Chinese children were well proportioned, but that as they aged their eyes began to appear smaller, and da Cruz added that if anyone in China did have *olhos grandes* they would have to be descendants of people from other nations. One of the paramount differences between antipodal Japan and Europe, said sixteenth-century missionary Luis Fróis, was that the Japanese hated large eyes and preferred their own, which were "closed at the tear ducts" (*fechados da parte dos lagrimais*).[3]

The notion of an epicanthus or "Mongolian fold," however, has more to do with eyelids than with the shape of the eye as such, and thus it is not surprising to find members of the medical profession as being among the first to attempt to identify it. The traveler François Bernier, whom we have discussed in chapter 2 as one of the first European taxonomers, was a physician by training, but his offensive description of "little pig's eyes, long and deep-set" was hardly medical in nature. Engelbert Kaempfer, however, who journeyed to Japan in 1690 as a physician with a Dutch embassy, briefly noted that the Japanese possessed "thick eye-lids," although, he added, "the eyes stand not so deep in the forehead as in the Chinese." In the early 1770s Carl Peter Thunberg also sailed to Japan as surgeon with the Dutch East India Company; he wrote that Japanese eyes were not only "oblong" and "small" and "sunk" ("squinting" in the original Swedish), but also that their "eye-lids form in the great angle of the eye a deep furrow, which makes the Japanese look as if they were sharp-sighted and distinguishes them from other nations."[4] In 1775 Blumenbach, also a doctor, proposed that East Asians had "blinking eyelids drawn up toward the external corners" (*oculorum palpebris conniventibus et in cantho externo sursum ductis*), later adding the French phrase *yeux bridés* (literally, "bridled eyes") by way of clarification.[5] The surgeon attending William Edward Parry's 1821–23 expedition to the Arctic noticed that among some Eskimos (by this time considered members of the Mongolian race, too) "the inner corner of the eye [was] entirely covered by a duplication of the adjacent loose skin of the eye-lids and nose." Simultaneously, another embassy surgeon to Siam and Vietnam mentioned "the tumid, incumbent eye-lid of the Chinese," and that "the aperture of the eye-lids, moderately linear in the Indo-Chinese nations and the Malays, is acutely so in the Chinese, bending upward at its exterior termination." Finally, in the early 1830s another physician-traveler to Japan, Philipp Franz von Siebold, seemed to bring the notion of a fold of skin covering the inner corner of the eye to a new level of precision. An entire chapter of his encyclopedic *Nippon*, published beginning in 1832, was devoted to a description of "the slanting of the eyes among the Japanese and several other peoples," accompanied by illustrations (figure 11).[6]

2. Erörterung des Schiefstehens der Augen bei den Japanern und einigen andern Völkerschaften.

Das Schiefstehen der Augen, welches man als ein bezeichnendes Merkmal in den Gesichtszügen der chinesischen Rasse aufgestellt hat, ist eigentlich nur ein Schiefstehen der Augenlider, ein Herabsinken derselben gegen die Nase. Es ist nicht zufällig[1], nicht gekünstelt[2], sondern eine im Bau der Schädel- und Gesichtsknochen dieses Volksschlages gegründete, eigentümliche Bildung der äußeren Teile der Augen.

Dieses scheinbare Schiefstehen der Augen, welches häufig zugleich mit einer auffallenden Kleinheit der Augenöffnung selbst vorkommt, beruht auf dem eigenen Bau des Stirnbeines und der Gesichtsknochen und auf einer daraus unmittelbar hervorgehenden Bildung der Augenlider.

Am Stirnbeine (os frontis) verliert sich bei diesen Völkern der Augenbrauenbogen (arcus supraciliaris) als ein weniger hervorstehender, aber breiterer Wulst in die Nasenfortsätze (processus nasalis ossis frontis), welche unterhalb der platten Glabella breiter und länger erscheinen, als sie bei der kaukasischen Rasse gefunden werden, und bei den Einschnitten (incisura nasalis) zur Aufnahme der Nasenbeine noch tiefer zurücksinken. Auch der Nasenfortsatz des Oberkiefers (processus nasalis ossjum maxillarium superiorum) ist mehr eingesunken, und es wird so die eingedrückte, platte Form der eben dadurch auch verkürzten Nase begründet.

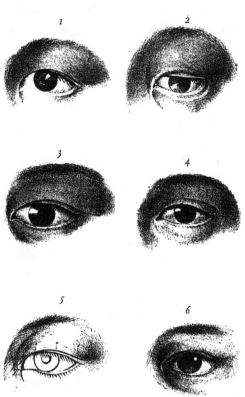

Fig 28. Vergleichende Tabelle der Augen bei den verschiedenen Rassen.

---

[1] Keine krankhafte Veränderung, wie Symblepharon, blepharoptosis u. dgl.

[2] Keine Verlängerung der Augenlider durch Zerren und Ziehen bewirkt, wie uns Buffon nach Gentil erzählt, noch andere absichtliche Entstellungen, wie unser würdiger Blumenbach geglaubt zu haben scheint. Histoire naturelle, Tom. VI, pag. 120, cinquième édition. — J. F. Blumenbachii, de generis humani varietate nativa. Göttingae 1711, pag. 48.

Figure 11: "Discussion of the slanting of the eyes among the Japanese and several other peoples," from Philipp Franz von Siebold, *Nippon* (1832) (1897 ed.). Slanted eyes, Siebold explains, are in reality a slanting of the eyelids that "sink" toward the nose. The illustrations depict (1) a Japanese eye, as well as examples from (2) Korea, (3) China, (4) Sulawesi, and (5) Borneo. Number 6 is a "normal" eye. Princeton University Library.

Siebold explained that "slanted" eyes were merely an effect of the lids "sinking" toward the nose that made the eyes appear higher at the outer corners. Due to the shape of East Asian skulls, he reasoned, with their "pushed-in" noses and their "protruding" cheekbones, the formation of the facial skin resulted in a "superfluous" fold that hung down from the upper eyelid and covered the inner corners. The smaller the nose the more "extra" skin there was, and the larger the cheekbones the more that fold of skin was forced downwards. This formation was not the result of disease or any sort of artificial modification, he insisted, alluding to recent European ophthalmology, which had just coined the term "epicanthus" but only with reference to a variety of conditions and eye diseases afflicting Europeans (figure 12).[7] He was also alluding to another old stereotype that East Asian mothers encouraged their children to pull continually at their eyes in order to elongate them.[8] But what was so different about his presentation was that "slanted" eyelids had for the first time become a marker of the race as a whole; in a later chapter he reiterated that they were one of the main "imprints" of the "Mongolian race" in general.[9]

The idea that East Asians had a distinct eyelid formation known as an epicanthus, in other words, had already been fully aligned with Blumenbach's Mongolian race by the 1830s. And in keeping with nineteenth-century scientific notions, this fold of skin also began to be connected to contemporary ideas about the relationship of the races to human evolution. As early as 1851 an ophthalmologist had proposed that the epicanthus, which seemed to occur more frequently among East Asians than other peoples, represented a "transition from the Caucasian race to the Mongolian race."[10] In so doing he was responding to new ideas in the field of natural science about the way that the races could be arranged according to ascending degrees of "perfection." An early proponent of this theory was Robert Chambers, whose *Vestiges of the Natural History of Creation* of 1844 remarked that "the leading characters... of the various races... are simply representations of particular stages of development of the highest or Caucasian type," just as the theory of recapitulation contended that the human brain progressed through evolutionary stages—from fish to reptile to mammal—before it could become fully

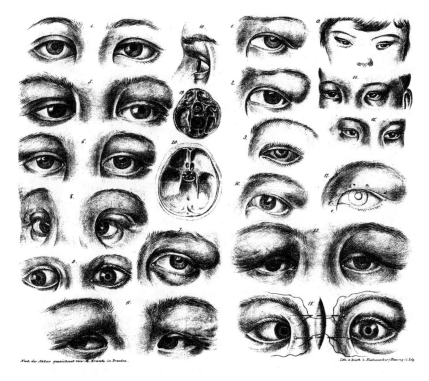

Figure 12: Varieties of epicanthus, from Friedrich August von Ammon, "Der Epicanthus und das Epiblepharon" (1860). Illustrations 1 through 12 represent different forms of epicanthi in European subjects; 13 shows a Siberian girl, 14–17 are copied from Siebold, 18 depicts an operation to "correct" an epicanthus, and 19 and 20 show the development of the epicanthus in European fetuses. Princeton University Library.

human. Further details were given by ethnologists and embryologists such as Etienne Serres, who in 1860 provided the example of the position of the navel, which, he argued, was highest among Caucasians, lowest among Ethiopians, and in between among in-between Mongolians—much as the navels of children were placed at a relatively low point in proportion to their bodies and only gradually developed into the adult position. Serres maintained, in other words, that Ethiopians represented not only a lower form of human life but indeed one that was equivalent to the infancy of Caucasians, and Mongolians, befitting their "childlike" intermediate nature, represented a slightly higher (or, rather, later) form of human development.[11]

And since it was also noted that an epicanthus could appear among European children and then fade away as the child grew, it only seemed to prove the point that as a "Mongolian" trait it was also something that a "higher" race would naturally pass through. It is in this context, moreover, that the phrase "Mongolian fold" was used for the first time in 1874—not in a medical journal but in the influential *Zeitschrift für Ethnologie*. Composed by the Russian immunologist Elie Metchnikov, it was a temporary departure from his earlier research in embryology and the result of fieldwork in Siberia among the Kalmuck people. He repeatedly emphasized what he called the "true Mongolian eye," as distinct not only from other East Asian variants but also from similar eye shapes that could occur among Westerners, and among Western children in particular. But his real aim was to demonstrate that this "true" Mongolian form, once again, was actually an example of an "arrested development" in the progress of human evolution, proving that Mongolians were in fact the oldest race in the world.[12]

Modern ophthalmologic classifications, on the other hand, divide epicanthic folds into four types, only one of which, the so-called *epicanthus tarsalis*, is particularly common among East Asians.[13] This was a complexity regularly felt within older scientific accounts as well, as "Mongolian fold," "epicanthus," or "Mongolian eye" were used interchangeably to describe anything from skin folds covering the canthus to much vaguer ideas about "narrow" or "slanted" eyes, or even the formation of a single as opposed to a double upper eyelid, which Siebold had also mentioned as distinct from a "slanted" eye fold.[14] It was variously described as a developmental "error," an "abundance" or "excess" of skin, and a hereditary "defect." By 1916 a leading anthropologist had tried to identify ten different types of fold that he further classified into subvariants, and in 1933 a certain grotesque climax was reached when it was proposed that the many types of fold could actually be organized ethnically and designated as "Mongolian," "Hottentot," "Indian," "Negro," and so on. Yet the one variety that temporarily affected European children, this author concluded, was normal and therefore had no scientific value.[15]

MONGOLIAN SPOTS

A somewhat different fate was in store for our second example, the "Mongolian spot," which by its very nature had a much more obvious connection to skin color. The term was coined at the end of the nineteenth century by a German physician working in Japan, Erwin Baelz (or Bälz), who in addition to teaching and operating a clinic at Tokyo Imperial University for many years, eventually served as personal physician to the Meiji emperor and his family. For Western readers Baelz became one of the great authorities on recently reopened Japan, and he was particularly influential for his theory, modified several times, that the Japanese were composed of three basic ethnic groups, both a "finer" and a "lower" type of Mongolian as well as the indigenous Ainu.[16]

It was in this context that in 1883 he published the first part of a long essay on Japanese anatomy in a German-language ethnology journal published in Tokyo. Among its many sections was a new treatment of the "Mongolian eye," which he, too, described as the result of "flat" noses and "excess" folds of skin. But his essay was to create the greatest stir in its description of Japanese skin color. After a standard description of the "light yellow" color of the Japanese and the "yellowish" color of East Asian peoples in general, Baelz paused to note a remarkable "dark blue spot that all newborn Japanese children bear in the sacral area or on their buttocks." After an extensive study of this mysterious phenomenon, he argued that it was caused by a particular kind of pigment deep beneath the skin, and that it began to appear in the fetus during the second half of pregnancy. The spot then faded away during the first year or two after the child was born. Might it, he wondered, also be found on children from parents of darker complexion in Europe? (figure 13).[17]

Much more than in the case of the epicanthic fold, it was supposedly clear from the start that the spot was a trait found only among Japanese and perhaps a manifestation of dark skin. The racial implications were obvious, too: even if, as Baelz also pointed out, people from the upper classes in Japan could be lighter than southern Europeans and southern Europeans could be just as "yellow" as the Japanese, truly white Europeans were not affected with any sort of color

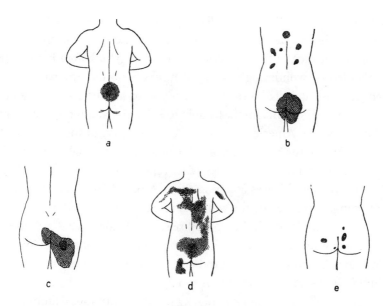

Fig. A.

Einige Beispiele für das verschiedenartige Auftreten der blauen Geburtsflecke beim Menschen.

a Japanisches Kind (n. GRIMM); b 10 Monate altes und c 3 jähriges Chinesenkind (n. MATIGNON); d Japanisches Kind (n. GRIMM); e 12 jähriger Knabe, Mischrasse von Grönland-Eskimos und Dänen (n. TREBITSCH).

Figure 13: Examples of "Mongolian spots," from Karl Toldt, "Über Hautzeichnung bei dichtbehaarten Säugetieren" (1913). The illustrations depict (a) and (d) Japanese children, (b) and (c) Chinese children, and (e) a Danish and Eskimo child from Greenland. Princeton University Library.

blot. What better proof could one hope to have that the Japanese were a nonwhite race? By way of confirmation a slightly later essay from 1900 confirmed that the spot could not be seen among the Ainu, a group that he and others had consistently categorized as "a Caucasian element" in the Japanese population. And by this time, too, Baelz was unabashedly calling the trait a *Mongolenfleck*, as it had also been seen in China as well as on half-Japanese half-European children, except for those who had "blonde fathers."[18]

His work was then given a far wider audience with an essay published in Germany in 1901. In addition to announcing plans for a detailed piece on the "Mongolian eye" that so far as I know never appeared, he insisted on the importance of the spot not only as an indicator of race but as an actual infusion of Mongolian blood. He concluded by wondering whether it might be found among American Indians (as it had already been reported among Eskimo babies in Greenland), as this might be able to solve the age-old question of whether Native Americans actually formed a part of the Mongolian race.[19] Six months later he reported seeing two native babies in North Vancouver with the spots, but as other reports began to pour in from around the world it was apparent that a Mongolian explanation was not going to suffice, for the spot had already been found in Indonesia, Vietnam, Thailand, the Philippines, and Madagascar.[20] Within a few more years it would also be seen among Hawaiians, Ceylonese, Central and South American Indians, and African Americans in the United States. And as one anthropologist asked as early as 1901, if the spots really were Mongolian, why were they not found among the people of Mongolia?[21]

Undaunted by this new evidence, however, Baelz became even more confident about the spots' usefulness as a racial marker, even as his rhetoric shifted away from their "Mongolianness" to the idea that they were utterly absent from white people. In 1901 he had called the spot "the most important differentiating marker between the Mongolian and the other races," but just one year later he asserted instead that it was "the finest reagent for differentiating the white race from all the others."[22] He had maintained this position, moreover, in spite of a serious challenge to his theory in 1903 by a Japanese professor of anatomy working in Europe, Buntaro Adachi, who had set out to discover whether it was true that the spot never appeared among Caucasians. After an extensive comparison of European skin samples he concluded that traces of the spot could be found in the white race as well, the difference being merely a quantitative one. Whether the spots were visible was dependent upon the amount of pigment present in different layers of the skin, but in any case the spot could no longer be called a racial marker. He provided a lengthy catalogue of the spot in Japanese medical literature dating back to the eighteenth century, where its existence had long been recognized and for which a variety of explanations—none of them racial—had been offered. References could also be found, he added, in far more ancient Chinese and Indian literature. Also included was a full list of Western accounts that dated back to 1816. It was not true that the spot failed to appear in the Ainu population, and he concluded with a brief report of a European child with the marks born of Moravian and German parents.[23]

But Baelz remained defiant; if the spot was *visible* only among certain races it was still a "tenacious" racial marker.[24] The Mongolian hypothesis may have been effectively discounted, but commentary on the spot was only just getting started. While Baelz himself was henceforth silent on the matter, over the next twenty years a veritable explosion of essays appeared in anthropological and medical journals. Alternative theories about the spots' source were regularly proposed. In 1901 a leading anthropologist hesitantly wondered whether it might be of Indonesian origin, and another commentator suggested African blood instead. In 1905 Albert Ashmead, former foreign medical director at Tokyo Hospital, agreed; "wherever you find black blood contaminating white," he wrote, "there you will find the mulberry spot of Japan." Others never quite abandoned the Mongolian hypothesis, such as one physician in Prague who explained the spots' existence on the Moravian-German child by pointing out that Moravia had been invaded by the Mongols in the thirteenth century. The 1911 edition of a treatise on tropical medicine concurred: if the spot appeared among the white race it was an "index of a distant effect of yellow blood, going back to the invasion of Europe by the Huns."[25]

Many of these commentators were far more racially virulent than Baelz, who had devoted his life to treating Japanese patients and who had married a Japanese woman. His own children, presumably, were "spotted" as well. And once scholarly commentary had shifted away from who had the spots and who didn't, as with any other purportedly racial characteristic the problem lay not in the identification of difference but in how that information should be interpreted and utilized. Befitting the scientific climate of the time, the spot was regularly treated as a sign of atavism in the evolutionary development of man, much as the "Mongolian eye" had been. An early report from Greenland in 1893, before the furor over the spots had even begun, had suggested this. When others joined the fray they became bolder about the details, such as the theory that in its original form the spot was the pigment that covered the entire body of the black races, but that when it was inherited by yellow people it would naturally disappear as the child "progressed" from black to yellow. Another reader suggested that the blotches of pigment were just "nuclei" of color that appeared first in the sacral area and later spread over the entire surface of the body, an idea that seemed to fit well with the observation that even black babies could be born white.[26]

In fact a similar interpretation had been offered in the earliest known Western record of all, composed in 1741 by a Jesuit missionary in South America. Some black children were born with black fingernails as a sign of what they would become, he noted, and the same was true for American Indian children, who were born with dark spots that faded as the babies lost their white color and became their "natural" one.[27] But Ashmead's suggestion was the most outrageous, that the spot was not only a sign of "negro descent" but indeed a "simian inheritance" and the trace of a monkey's tail—the implication being, of course, that anyone with nonwhite blood was far closer to a monkey than someone without it. Even as those with "Mongolian eyes" had too much skin, those with the spot had "too much pigment in the blood," and "the child of such parentage cannot get rid of its excess before birth." In a separate essay on syphilis in Japan, Ashmead characterized the entire nation as engaged in nonstop promiscuity and prostitution, pederasty, concubinage, public bathing, and infection—all of this supposedly relating to the question of "the capacity for development in the Japanese race."[28]

Much as Baelz had argued, for many interpreters the spot had become more of a marker of nonwhiteness than of any particular race, as if these children had simply been dipped into an inkwell that had stained them, however temporarily, as members of a "colored" group. It was no accident that one of the early names given to the spots was "congenital stigmata." In 1928 a physician noted as many as seven competing theories to explain its incidence in different racial groups, and it also found its way into the medical anthropology that formed the basis for extremely treacherous programs of so-called racial hygiene. Such standard texts as Baur, Fischer, and Lenz's 1921 *Menschliche Erblichkeitslehre* (The Study of Human Heredity) included a section on the spot in the context of a "Mongolian influence" on (Aryan) Europeans, who of course were precisely the ones seen as being in need of protection. When the tome grew to twice its original size in 1936, the presentation of the *Mongolenfleck*, like the rest of the volume, had become a form of scientific racism featuring mathematical symbols of hereditary descent and a genetic formula to account for its greater visibility in the "dark" races. The purpose of this was to emphasize the dangers of European "cross-breeding."[29]

In subsequent decades suggestions about their ethnological significance were occasionally put forward. One contention was that they revealed the origins of Native Americans; another was that they proved that the people of Madagascar were of Indo-Oceanic descent and linked to other black peoples of the world. But for the most part serious anthropological interest in the spots seemed to wane, even as the reasons for their occurrence and disappearance remained unclear.[30] It is interesting, however, that the term itself was far more persistent than "Mongolian eye" or "Mongolian fold," perhaps because a "Mongolian spot" was considered less of a racial slur, since this particular trait was so transitory and, to a large degree, hidden. Yet it did find its way into standard pediatrics textbooks where it yet remains, although often in quotation marks. A number of modern physicians also began to note cases where the spots were mistaken for child abuse since they look so much like bruises; many Western doctors, it seemed, were as unaware of their existence on "nonwhite" children as Baelz had been a century before.[31]

"MONGOLISM"

Our last example is "Mongolism," which in terms of its coinage is actually the earliest of the three, having been first proposed by the English physician John Langdon Down in 1866, and in honor of whom the disease is still generally known today as Down syndrome. The study of Down syndrome sprang from very modest beginnings: a report of only four pages published in a London hospital journal and titled "Observations on an Ethnic Classification of Idiots."[32] "Idiot" was one of the names by which mentally disabled people were known in the nineteenth and early twentieth centuries, in a system of classification that also included "imbeciles" and the "feeble-minded." Down was one of the first physicians to devote his time and energy to improving the treatment of these individuals, and he struggled, as other physicians of the day were to do, with a means of grouping together those placed under his care for the purpose of understanding how best to treat them. Some twenty years later, for instance, it was proposed that a more useful scheme would be to divide congenital conditions from those that were either acquired or accidental, but Down's suggestion was to organize them around what he called "an ethnic classification."[33]

In an excellent brief introduction to Down and his scientific background, Stephen Jay Gould has pointed out some of the ways in which even the terminology chosen was affected by widespread prejudices of the day, especially Down's strange decision to group his patients according to a racial classification, given that all of them would have been white. This was because a large number, Down argued, resembled "one of the great divisions of the human family other than the class from which they have sprung." Some, he elaborated, looked like Ethiopians with "prominent eyes," "puffy lips," and "woolly hair"; others seemed to be Malay with "soft, black, curly hair" and "capacious mouths"; still others appeared American with their "prominent cheeks" and "slightly apish nose." Lastly were "typical Mongols," with "flat and broad" faces, "obliquely placed" eyes, epicanthi, and skin with a "slight dirty yellowish tinge." The members of this last group, he added, seemed to share such a close resemblance that "it is difficult to believe that the specimens compared are not children of the same parents."[34]

These four racial types, of course, were exactly those established by Blumenbach in 1795, and while Down's other labels were not taken up by subsequent physicians, his delineation of a "Mongol" type became standard, and these patients were henceforth called "Mongoloids." His choice of the term "Mongoloid" rather than "Mongolian" was primarily an English-language phenomenon and probably due to the contemporary influence of Thomas Huxley, who regularly used it; there was a different suggestion to use "Kalmuck idiocy" instead, but for some reason this was deemed more objectionable than "Mongol" and few adopted it.[35] Once again, moreover, we can see a preoccupation with ethnicity based on the same "ladder approach" to race that sought to explain the existence of Mongolian eyes and Mongolian spots. In other words, for Down this was not merely a question of resemblance but a "natural system" of "degeneration" and "retrogression." Chambers's theory that East Asians represented "the child race of mankind" seemed to take on a whole new significance, since white "Mongoloid" patients were envisioned as a kind of throwback to racial Mongolians, whom Chambers had also characterized as "arrested infant[s] newly born."

And as fuller descriptions of "Mongolism" became available its sufferers were said to share other stereotypical East Asian traits such as a "childlike" language and a predisposition toward mimicry. Chambers was hardly alone in calling the Chinese tongue "destitute of all grammatical forms" and "a language which resembles that of children, or deaf and dumb people." And the idea that the inhabitants of the East were mere imitators rather than creators had an equally long pedigree.[36] One trait that "Mongoloids" and racial Mongolians did not seem to share, however, was the Mongolian spot, typically seen as a feature of "real" Mongolians only.

Down's interest in "ethnic classification" was part of a general trend to quantify all sorts of mental variation not just in biological or social terms but as racialized "regressions" from Caucasians. Criminal anthropology, to take another example, attempted to find evidence of nonwhite "simian" inheritances in those who were judged to be criminally insane, and Down's racial comparisons of asylum patients should be seen in a similar light.[37] Yet race was also of interest for Down because of his explicit desire to reinforce the notion of monogenesis in the aftermath of the American Civil War; if forms of "degeneracy" were traceable in "Mongoloids" and other "idiots," he argued, all the races must stem from a single source, including African slaves. Yet we also have to remember that the focus was on racial questions as points of comparison, and that the objects of research were European whites who only seemed to resemble the Mongolian race, among others.

Later researchers, though, repeatedly emphasized that this resemblance was merely superficial, one of the earliest dissenters being Down's own son in 1906, although he still implied that there must be some sort of "reversion" at work. Other early authorities noted a dissimilarity in terms of skin color; "Mongoloid" skin, wrote G. E. Shuttleworth, was "sallow," not "yellow" like "the genuine Mongol[ian] infant." And by the 1940s even the shape of the eye, once considered the most "Mongolian" of all Down syndrome traits, was said to be distinguishable from the epicanthic formation most common to East Asians.[38]

Similarly, it was not until the first half of the twentieth century that it was even recognized that Down syndrome was not a condition peculiar to Caucasians, a fact that would naturally represent a significant blow to the theory that it was a sign of retrogression. Much as in the case of "Mongolian spots," it seemed that scientific ignorance was due largely to the fact that nonwhite subjects had simply been excluded from consideration. But by the 1920s a few cases of "Ethiopian mongolism" and (most ludicrously) "Mongolian mongolism" were being reported, although many scholars remained convinced that the disease must have some sort of anthropological significance.[39] One of the most notorious was Dr. F. G. Crookshank, a fellow of the Royal College of Physicians and author of numerous papers on topics ranging from migraine and influenza to venereal disease and flatulence. He was best known as a maverick leader in attempts to combine medicine and psychology, particularly the individual psychology of Alfred Adler, along with a strong dose of eugenics, literary modernism, and Nietzschean philosophy of the will, all much in vogue during the interwar years in Europe. In 1924 he created a sensation with his *The Mongol in Our Midst: A Study of Man and His Three Faces*, a short novelistic treatment of just over one hundred pages in which he took atavistic ideas about "Mongolism" to a whole new level.[40]

He agreed that Down syndrome, which he associated with "white" people only, was an instance of retrogression, but he also argued that each of the "three faces" of humanity (Caucasoid, Negroid, and

Mongoloid) embodied an array of correspondences that linked them to the three varieties of anthropoid ape: the chimpanzee, the orangutan, and the gorilla. Mongolians were "orangoid." The three-ape thesis was not new; one of its earliest proponents was zoologist Carl Vogt in 1863, and Darwin had criticized it in *The Descent of Man*. But by Crookshank's time it was being revived, especially by the physical anthropologist Hermann Klaatsch, who after studying the remains of a prehistoric skeleton found in southern France in 1909 concluded that its bones were entirely different from those of Neanderthal man, and that perhaps early man should be divided between a western group and an eastern one, which he termed "gorilloid" and "orangoid," respectively. Indeed the two types might represent different species, their differences extending even to the brain.[41]

Such an apparent return to polygenism (or polyphyleticism) did not go unchallenged, and Klaatsch's following seemed limited.[42] But in 1913 one disciple by the name of Eugen Kurz published a paper from Shanghai in which he consistently differentiated "the Chinese brain" from the European one, concluding that while the former had progressed from the level of the apes it was still far smaller than its Western counterpart. In a series of subsequent studies he repeatedly contended that Chinese bodies were both fundamentally "primitive" and entirely distinct from self-described European norms, and in 1924 his fullest statement on the subject determined that the "yellow race" represented an entirely separate branch of humanity that, in short, had "sprouted from the same root" as the orangutan.[43]

Crookshank did not mention Kurz's work at first, yet when *The Mongol in Our Midst* went into a second edition in 1925, Kurz was credited with having "establish[ed] in the most convincing manner the strictest homology between the Chinese and the orang brains." The love affair was mutual: Kurz became Crookshank's German translator in 1928, and in Crookshank's third edition Kurz was one of the most frequently cited authors.[44] Yet what set Crookshank apart from his German colleagues was the way he allied their notions of a "primitive" "Eastern" "orangoid" man to Down's concept of the ethnic classification of "Mongoloid idiots." The result was a grand edifice of absurd allegations about the orangutan, the racial Mongolian, and the "Mongoloid," all of whom were said to share a variety of homologies that demanded further investigation.

It was no coincidence, he alleged, that they all instinctively sat in a cross-legged "Buddha-like" position (never mind that Buddhism was an Indian import), or that they all shared other parallels such as hand-markings, the shape of the face, the texture of the skin, and even male and female sexual organs. Epicanthic folds, too, were supposedly evident in all three, but the spot, which did not seem to appear among "Mongoloids" at all, was simply discounted. Could all of this prove the validity of the theory of recapitulation upon which Down's original idea was based, or that the white race had developed from the "lower" ones, or that those with Down syndrome represented a "Mongol stage" of human evolution? Might this be the result of genetic influence, proving that in Europe's distant past there had been a substantial infusion of Mongolian blood? How else might one explain the "range of Mongolian or semi-Mongolian types amongst our native Cockneys"? (figure 14).[45]

The book created what one reviewer called "a pleasant furor, especially among laymen," although responses from the scientific and medical communities were much more muted. But they, too, often seemed to support at least some of Crookshank's assertions, particularly about the reality of the "lower races." A reviewer in the 1925 *Archives of Pediatrics* objected to the idea that Down syndrome was proof of Mongolian blood and explained patients' cross-legged posture as a natural result of the looseness of their leg and hip joints, yet this reviewer still agreed that "the individual in his embryonic development recapitulates the phases through which he passed in the history of the race in its evolution from lower to higher forms." Some saw through the claptrap at once, among them the geneticist Lionel Penrose, who would soon begin groundbreaking studies on Down syndrome, demonstrating that its source was not racial at all. Another was American linguist Edward Sapir, whose scathing review in *The Nation* briefly wondered whether the whole thing might be a hoax. "The author's thesis," he remarked, "need only be stated to be refuted with a laugh."[46]

Fig. 1. A "MONGOL" SCHOOLGIRL.
(London.)

Fig. 2. A RACIAL MONGOL.
(Mongolia.)

Fig. 1. AN "ASYLUM MONGOL."
(Dr. Shuttleworth's Case).

Fig 2. A KIRGHIZ MAN.
(Racial Mongol).

Figure 14: Four illustrations from F. G. Crookshank, *The Mongol in Our Midst*, 3rd ed. (1931). The photographs were supposed to show the physical similarities between "racial Mongols" and "asylum Mongols," i.e., Europeans with Down syndrome. Princeton University Library.

Crookshank, however, was deadly serious, and not content with his moment of fame, he felt called upon to respond to his critics by expanding and entirely rewriting the book, which when it was republished in 1931 had grown to more than five hundred pages. Weighty, ponderous, and hugely overstated, each of its sections had now been fleshed out with new and supposedly scientifically validated evidence, including extensive material on "the Chinese brain." He also began to wonder whether there might be some kind of mystical "homing instinct" that drew "Mongols expatriate" toward the East or toward anything associated with it. Examples are given of people with "Mongolian physiques" who had an otherwise inexplicable interest in the Orient. If a long-term British resident in China came home "looking like a Chinaman," Crookshank mused, it may have been that "he went there because he was one, and stayed there because the Chinese thought so, too."[47]

The suggestiveness of such claims obviously intrigued many readers, and "Mongol" became something of a stylish catchword, interchangeable with "Oriental" and "Asiatic." But what strikes a reader today is the manner in which Crookshank's racial stereotypes were rarely questioned. Perhaps the homologies were nonsense, but few seemed to object to describing Down syndrome children as behaving like "little Chinese mandarins," nodding and swaying "in the very act and manner of the kow-tow" or making "the same errors as does the Chinese who speaks pidgin English." Whether or not anyone agreed with Crookshank, they did not seem to dispute his characterization of the Mongolian race as infantile, playful, soft, Buddha-shaped, flat-faced, and imitative—and of course yellow: "some of the imbeciles are as sallow and earthy . . . and definitely yellow as are many Chinese or Japanese."[48]

It was not until the end of the 1950s that new knowledge about the chromosomal nature of the disease would lead to an increased awareness of its racialized labels. Pleas to change the name were printed in medical journals in 1961, and by 1965 the government of Mongolia had lodged a similar protest with the World Health Organization. Yet at the time many researchers and clinicians remained hesitant. The published proceedings of a 1966 conference to commemorate the centenary of Down's original contribution was still titled *Mongolism*, and a Japanese participant's reminder to those present that they had "a responsibility to resolve this technical problem" was met with little support. And yet this was an issue involving much more than the simple choice of a name: a major 1978 treatment defended the use of Mongolism in its subtitle by noting that "Down's Syndrome children do, after all, look 'Mongolian.'"[49]

In all the many debates over Down syndrome terminology, that is, there is little record of anyone who actually challenged the notion that all East Asians *were* Mongolian, racially speaking. Similarly, in the case of Mongolian eyes and Mongolian spots, "Mongolianness" itself was never at issue. If a Westerner "looked like" a "Mongol" because of Down syndrome or because of a particular shape of the eye or a mark on the skin, these features were characterized as distinctly foreign or exotic elements in the Caucasian physiognomy.[50] Theories of retrogression, too, may have been eventually abandoned, but there was still no question that a Western body with such anomalies had to be explained, or, at the very least, apologized for.

"MONGOLIAN" BODIES

Despite all its eccentricities, one of the most familiar things about Crookshank's reading of Mongolism is his breathing new life into the centuries-old notion that if there were "Mongols" in "our" midst it was because the Mongolian was an essentially invading race. Like epicanthic folds, pigment spots, imitative brains, and childlike physiognomies, these were long-standing prejudices about East Asian bodies that found their way into medical characterizations of "Mongolian" conditions. And much as in the obsessive quantification of skulls, bones, faces, and skin color that preoccupied the study of comparative anthropology, medical researchers and practitioners were obsessively trying to measure white Western normalcy as against the physiological and pathological "defects" of the Mongolian.[51] A sacral pigment spot or an epicanthus could be seen as debilitating markers of "Mongolianness," and Down syndrome, which was originally thought to afflict white bodies only, was viewed as a similar taint of a "lower" race.

Yet also like the anthropological debates we have reviewed in the previous chapter, medical deliberations regarding Mongolian eyes or Mongolian spots or Mongolism were not just derived from contemporary theories about race but actually helped to produce and to strengthen them; "Mongolianness" was clearly a rationale for racism just as much as the other way around. An excellent case in point is Chambers's explanation of ethnic hierarchies, which could supposedly show that there really were measurable differences between the races, a circular process of reasoning that was made more difficult when (as was typical) the data failed to demonstrate what the researcher had set out to prove. This is nowhere more apparent than in the case of skin color, which Chambers, interestingly enough, also claimed to be an example of racial "perfection." "Why are the Africans black," he asked, "why are the Mongolians generally yellow, the Americans red, the Caucasians white? . . . All of these phenomena appear, in a word, to be explicable on the ground of *development*. . . . May not colour, then, depend upon development also?"[52]

But as I have repeatedly asked, how might one actually be able to explain yellow as a "developmental" color? Why should one decide upon yellow rather than an infinite array of other possibilities, many of which were chosen before yellow (or Indian red) and had become a collective Western fantasy? How does one explain yellowness on the basis of theories of arrested development, recapitulation, or degeneration? The Mongolian eye fold might be fetal in all races; Mongolian spots were often described as a racial throwback or a trace of pigment that either disappeared or spread out over the entire surface of the body; and Mongolism was said to bear a certain resemblance to a fetal formation as well as to a "lower" race. But the yellowness of the Mongolian was simply assumed from the start, although a variety of natural metaphors (dead leaves, desiccated lemon peel, jaundice) were regularly invoked to support it. In Chambers's case it was not that East Asians simply looked yellow or that yellow was the most appropriate intermediate shade between black and white. Rather, and in a completely backward way, since the theory of recapitulation and the hierarchy of races were already accepted facts, then yellow must be the most appropriate developmental shade. His only explanation, such as it was, was that at any stage of human development the skin might be "predisposed to a particular colour."[53]

We may no longer refer (at least in English) to East Asians as Mongolians or to Down syndrome as Mongolism, but Mongoloid is still regularly used in certain scholarly fields along with Caucasoid and Negroid, and Blumenbach's placement of East Asians as antithetical to the West is still being felt after more than two hundred years. Moreover, the idea that these people are yellow has lost little of its hold. Some anthropologists had set out to prove by means of a color top that East Asians were yellow in fact, but physicians such as Down did not seem to need any convincing. The skin of "Mongol" patients, he wrote, was of "a slight dirty yellowish tinge" just like real East Asians. Baelz similarly remarked that Japanese skin was "of a light yellow color," and in one of the earliest studies of "the Chinese brain" published in 1891, the cadaver that the anatomist was beginning to dissect was matter-of-factly described as having "the characteristic yellow complexion."[54] One would have little difficulty multiplying examples of this view, and indeed it would be no easy task to find many dissenters. Even Gould's account of Down's racial prejudice, after remarking that children with the syndrome did not really look East Asian, still added that some of them "have a small but perceptible epicanthic fold, the characteristic feature of an oriental eye, and some have slightly yellowish skin."[55]

I do not mean to denigrate Gould's otherwise excellent critique, nor his fundamental investigation of the history of the theory of recapitulation (among other subjects), but his offhand characterization of "oriental features," including yellowness, remains much in the spirit of the medical "Mongolianness" we have been tracing in this chapter. As in late-nineteenth- and early-twentieth-century anthropology, the medical explanations for "Mongolian" conditions had an uncanny way of reinforcing the stereotypes with which researchers began. And in terms of yellowness, we will see in the next chapter that it would be further strengthened—and indeed solidified—when it was taken up at exactly the same time by East Asian cultures themselves.

Chapter 5
Yellow Peril

The Threat of a "Mongolian" Far East, 1895–1920

In previous chapters I have repeatedly insisted on the difficulty of determining any sort of moment at which East Asians had suddenly "become" yellow. During their initial encounters with the West they were almost uniformly described as white, and while they slowly darkened in European eyes there was never really a consensus about exactly what color they were. A defining moment occurred at the end of the eighteenth century when they were lumped together into a new racial category called the "Mongolian," commonly identified as the "yellow race." But even then a number of color terms continued to be used. Late nineteenth- and early twentieth-century science took up the question of "Mongolianness" and sought to quantify it in a number of ways, the result being that yellow skin seemed to become an accepted fact of East Asian bodies. Yet it was only at the end of the nineteenth century that the idea of a yellow East Asia would fully take hold in the Western imagination, crystallizing in the phrase "the yellow peril" to characterize the perceived threat that the people of the Far East were now said to embody. Not only did the discourse of yellow become ubiquitous in the West at this time, but we will also see in this chapter that it migrated into East Asian cultures, provoking a range of responses, reactions, and incorporations. Yet before we begin to examine some of this material we should note first how revealing it is of Western Orientalist attitudes that yellow skin would become fully agreed upon only when it was accepted as a feature of the "perilousness" of the region, thus firmly bringing together what had been closely allied for centuries: yellow skin, numerous "Mongolian" invasions, and the specter of large numbers of people from the region migrating to the West.[1] Previous to this moment of perceived crisis, if Westerners had any difficulty finding words to describe the color of the inhabitants of China or Japan it was not because of any sort of fear or anxiety about getting it right. I see it as more of a blithe disinterest in something that did not matter much either way, since, after all, these people were not really white and since, after all, they remained in their own country. But by the end of the nineteenth century substantial numbers of East Asian immigrants actually began to appear on Western shores and in Western communities, and the need to racialize had become stronger than ever.

We have seen that fears of a "barbaric" East had been linked to danger in the Western imagination for centuries, although the names given to the invaders varied considerably: Persians, Scythians, Parthians, Huns, Turks, Tartars, Mongols. By the eighteenth century, just before Blumenbach's classification of the Mongolian race, Gibbon had included a special chapter on the uncivilized northeastern "pastoral nations" as one of the main causes of the collapse of the Roman Empire. In 1819 William Lawrence described the "Mongolian tribes" as "like the deluge, the tornado, and the hurricane," and in 1848 S. Wells Williams's standard study of China spoke of the "vast swarms" of Mongols that had once "overrun the plains of India, China, Syria, Egypt, and Eastern Europe." By the 1870s China was being called a "yellow terror" (*gelbe Schrecken*), and in 1877 anthropologist Armand de Quatrefages provided a map to show that the Mongolian *races jaunes* comprised 44% of the entire world's population. In 1880 Gobineau even warned of "a new fifth century" if an "Oriental movement into Europe" was allowed to occur; his last work of 1887, published only after his death, was an epic account of such a battle between the "whites" and the "yellows."[2]

But with the invention of "the yellow peril" in 1895, the notion that East Asians were yellow *and* perilous had become ubiquitous, not only because this phrase appeared in every imaginable category of both scholarly and popular writing, but also because for the first time it was able to cross European linguistic boundaries. East Asians were now definitively yellow because they were a *péril jaune*, a *pericolo giallo*, or a *gelbe Gefahr*. A number of modern studies have begun to outline the tremendous complexity of these terms; one of the best accounts remains Richard Austin Thompson's 1957 Ph.D. dissertation, which brilliantly demonstrates that the peril embodied multifarious anxieties not simply about the prospect of increased East Asian immigration, but also of military aggression, economic competition, and social degeneration. The objects of these fears were similarly free-floating, just as in the popular imagination the supposedly perilous qualities attributed to Chinese or Japanese frequently seemed to trade places, with one group being lauded even as the other was anathematized as the true enemy.[3]

For instance, in the middle of the nineteenth century increasing numbers of Chinese workers had come to places such as the American West, Canada, and Australia, to mine gold or to build railroads, and they were despised in part because they did the same jobs as their Western counterparts but did them better and more cheaply. In 1895 the Sino-Japanese War (also known as "The Yellow War") aroused fears about Japan's aspirations as an imperial power, a tiny country that had been officially open to outside influence for only forty years and had already defeated its vastly larger and more populous neighbor. The Boxer Rebellion of 1900 provoked anxiety about the potential explosiveness of millions of Chinese in a crumbling Qing empire. The victory of Japan at the conclusion of the Russo-Japanese War in 1905, finally, seemed to mark the end of Western control of the "civilized" world.

The most famous icon of these fears was a widely distributed 1895 engraving made after an original drawing by Kaiser Wilhelm II of Germany (figure 15). It is not entirely clear when the title was added, but it soon acquired the label *The Yellow Peril*, and the phrase quickly became a catchword for the eastern "situation."[4] As an allegory it could hardly be more apparent that the Far East, here represented by a statue of Buddha sitting atop gigantic clouds of fire and smoke, was perceived as a source of impending danger. At the left of the image, overlooking the immense (European) valley that lies between them and the fiery destruction in the distance, stands a group of warrior women representing the nations of Europe, being called to arms by

Figure 15: "The Yellow Peril," from *Harper's Weekly* (1898). An engraving made after an original drawing by Kaiser Wilhelm II, with maidens representing the nations of Europe being called to arms under a Christian cross to combat the threat from the East: a Buddha seated atop huge clouds of fire and smoke. The image is captioned in Wilhelm's handwriting at the bottom center, with accompanying translations: "Nations of Europe! Join in the Defence of Your Faith and Your Homes!" Princeton University Library.

the archangel Michael with a flaming sword. The picture was also signed with the Kaiser's handwritten inscription: "Völker Europas wahrt eure heiligsten Güter!" (Nations of Europe, defend your holiest possessions!). A blatantly religious allegory as well, a large Christian cross appears in the sky above the European maidens, to be distinguished, of course, from what an authorized explication published in the German press called the "heathen idol" Buddha, which, we also read, sits atop a "demon of destruction," a Chinese dragon (difficult to make out in the image, which seems to show just a thundercloud). In letters to the Russian tsar, Wilhelm spoke of defending Europe from "the inroads of the Great Yellow race," "the inroads of the Mongols and Buddhism," and "the inroad[s] of Buddhism, heathenism, and barbarism," all of which were characteristically equated.[5]

A number of Western commentators immediately noticed how ironic it was that Buddha, the embodiment of self-effacing nonviolence and pity par excellence, should have become the agent of such destruction, but this was typical of many Europeans' willful blindness towards the cultures and religious beliefs of other nations. Wilhelm himself was obsessed with the image and referred to it constantly during the remaining years of his life, particularly after Japan's defeat of Russia in 1905; "remember my picture," he wrote to the tsar two years later, "it's coming true!" Copies were placed in German steamships traveling to the Far East, and the Kaiser even had postcards printed up that he continued to use in his correspondence for many years afterward.[6]

THE RECEPTION OF YELLOW IN CHINA

But how were these ideas received in China and Japan? Over the course of this book I have argued that Westerners' choice of yellow was not based on travel texts or missionary reports about the region, nor on any sort of knowledge about the importance of the color within the Chinese tradition. Instead, it was dependent on numerous historical, scientific, and cultural contingencies that had little to do with the "real" Far East. Its reception in East Asia, however, was anything but accidental, and in China in particular, where the Western idea of a yellow (but not "Mongolian") race was frequently perceived as an *appropriate* color, and it became adopted by many in China as a form of wholly positive self-identification. This was not merely an appropriation of yellowness but of Western racialism itself, particularly its strange preoccupation with determining a particular color for each geographical group, a mode of thinking for which there was no real precedent in Chinese (or indeed any East Asian) thought. Certainly East Asian traditions, too, had made distinctions about skin color according to gender or to social rank, "white" upper-class beauty as opposed to those who were compelled to labor or to spend time outdoors, but this was hardly the same thing as the Western taxonomical and anatomical traditions that, along with many other aspects of Western science, were being introduced into China and Japan beginning in the second half of the nineteenth century.

In the case of China, the end of the Qing empire led many reform-minded thinkers to consider importing a variety of ideas from the industrialized West, including its scientific notions of biology, heredity, and evolution.[7] Discussions of Western racial categories also became common, although often accompanied by considerable readjustment. It was frequently the case, for example, that while these thinkers accepted the fact that Chinese were part of the "yellow race" they did not adopt Western ideas about white European supremacy. Kang Youwei's grand utopian vision in the *Da tongshu*, or *Book of the Great Union* (ca. 1902), for one, suggested that a "great union" between the whites and the yellows would result in a new world order based on Confucian principles. He agreed that the world was divided into races of different colors, and that the "dark" races were inferior, but instead of a simple privileging of the white race he adjusted the hierarchy to put the "yellows" on the same level.[8]

Such appeals within China were often accompanied by a reification of the Yellow Emperor, or Huangdi, the supposed founder of the "pure yellow" or Han race, even though *huang* here is not necessarily a color at all (it is also a family name, and the character embodies many other symbolic meanings). The phrase "Sons of the Yellow Emperor," actually "Sons of the *Yan* (Fire, Red) and *Huang* (Yellow) Emperors," still in use today as a descriptor of the Chinese diaspora, could have considerable political utility during the late Qing and early Republican periods, as the concept of yellow could be used to argue that the "real" China, as represented by the Han people and the mythical Yellow Emperor, needed to be liberated from the "Mongol" or "foreign" Qing rulers.[9] A related example was the foundational text of Chinese medicine, the "Yellow Emperor's Canon of Internal Medicine" (*Huangdi neijing*), sometimes shortened to the phrase "Yellow Canon" to differentiate it from the Western one.

This was hardly the first time in Chinese history that those in power had sought a form of legitimacy by invoking the Yellow Emperor as a cultural, scientific, or political ancestor, although architects of the Republic such as Sun Yat-sen did present a more integrated vision of China's multiethnic population. Yet his ideas were also firmly based on Western racialized categories. "Mankind is divided first into the five main races," he wrote in his foundational *Three Principles of the People*, "white, black, red, yellow, brown.... Chinese belong to the yellow race because they come from the blood stock of the yellow race." But once again a distinctly different color coding could result, as in the first flag of the Republic, known as the "Five Races Under One Union" and in use until 1928. Consisting of five horizontal stripes (red, yellow, blue, white, and black), each color was supposed to represent one of the ethnic groups now unified under the new Republican government. Red represented the Han people, while yellow was chosen for the Manchus.[10]

These examples were not just passive acceptances of the notion of yellow as it was understood by the West, even if many in China seemed to agree that they were, in fact, members of "the yellow race." In Kang's book the Chinese were also called "gold-colored people," conceptually superior, we would assume, to the "silver-colored people" of the West.[11] Some commentators simply inverted the term "yellow peril" as an image of Chinese glory and superiority, often conveniently overlooking the fact that Mongol invasions had constituted a peril for China as well. Others retorted that the real "yellow peril" was Russia, or that the West as a whole, with its heinous imperial policies, was nothing more than a "white peril" for the rest of the world, a view with which a number of antiwar or anti-imperialist Westerners agreed.[12]

What is so fascinating about these developments is that a different kind of yellow that had long existed in the Chinese tradition was now being reinvoked. The Chinese could appropriate the Western term because it fit preexisting myths about their own civilization. In China yellow was the color of the earth and the central color, the official color of the Emperor and the color of the muddy Yellow River considered to be the cradle of Chinese civilization. But it had never been perceived in China as the color of Chinese skin. There were certainly suggestions about yellowness in Chinese physiognomic writing, as in the phrase "yellow between the eyebrows," which is a color signifying happiness or good fortune, perhaps related to the yellowness of the face and the robe of the Sakyamuni or historical Buddha. There is also a Chinese creation myth referring to the first human beings as "yellow soil people" (*huangtu ren*), since the goddess Nuwa was said to have created them out of the earth.[13]

Pre-nineteenth-century Western readers would have been familiar with many of these details, as they had become routine once they were introduced by seventeenth-century Jesuit narrators like Matteo Ricci and Martino Martini. The story of Nuwa was also explained in the encyclopedic Jesuit *Mémoires* in 1776.[14] Yet there is absolutely no evidence that any of this information had any direct influence on the initial racialization of all East Asians as a "yellow" people. At the conclusion of chapter 2 we mentioned that Blumenbach had used the tempting phrase "yellow man of the East" as a nickname for a Chinese visitor to England in the same decade. But even Blumenbach, the inventor of the five-part racial scheme that still largely holds sway today, failed to make any connection between this nickname and his yellow "Mongolian" race, which, after all, was based on a separate taxonomic tradition in which notions of color were drawn from medicine, botany, and comparative anatomy, not from travel or missionary texts.

Another example from the same decade is the French magistrate and scholar Charles de Brosses, author of a 1765 linguistic treatise on the "mechanical formation" of word etymologies. He followed it with a long specialized essay in 1774 on the variety of names given to the Scythians, one of the many designations used to describe the peoples of Northeast Asia before "Mongol" had been popularized. Constructing a complex set of claims surrounding the meaning of such names as Mongol, Kalmuck, and Tartar, de Brosses eventually came to the famous classical terms "Seres" and "Serica," used by Strabo, Pliny the Elder, and others to designate a people and a land in Northeast Asia that almost certainly corresponds to some part of present-day China.

As de Brosses realized, these words had always been identified with silk: the people of Serica were "the silk-wearing people," and the Latin adjective *sericus* meant "silken." But he also argued that for "the Tartar nations" *seres* was a name for yellow (the Mongol word is *sira*, *ser* in Tibetan), and that this was the real origin of the term *sericus*. It is very simple, he commented, that "dressed in this color and trading in this precious merchandise they should have received from strangers the name of *Seres*, i.e. *peuple jaune*." The etymology of "Seres" is still a subject of some dispute, but it seems clear that de Brosses was adding what for an eighteenth-century reader had become a standard detail: that yellow was of great importance in China. But to claim that the Chinese were the "yellow people" because they wore silk (because raw silk is of a yellowish color?), or because that color was especially prized in Chinese culture, was hardly the same as claiming that they are members of a "yellow race," which of course neither de Brosses nor anyone else could do in the 1770s.[15]

A third example occurred in a three-hundred page dissertation on Chinese history by missionary Joseph-Marie Amiot, also appearing in the Jesuit *Memoires*, which attempted to organize all of Confucian learning into a single system illustrated with a series of charts. Among the many conceptual categories that Amiot presented for European readers was the notion of the five colors, blue, white, yellow, black, and red, all of which have numerous correspondences and significations in Chinese scholarly and religious traditions, including divination, astrology, and medicine. The only comment Amiot provided, however, was that the colors could be assigned to human beings as well, and that "judging from the ancient paintings that one can see here [in China], they really did have people of different colors; even today, I might add, it is not impossible to find such people, above all in the southern provinces of the empire."[16]

The question of Amiot's Chinese sources is a difficult one, but there can be little doubt that he was offering a peculiarly Western understanding of the idea of the colors as applied to variations in human complexion, which had little place in Chinese thinking, although colors could also be assigned to each of the "Five Elements" or *Wu xing*, which he also mentioned in the same section. We have seen in chapter 1 that claims regarding the relative darkness of the southern Chinese was a long-term preoccupation for Western visitors, yet it is also revealing to see how Amiot's commentary was received more than one hundred years later, in the context of very different ideas about human color. For in 1885 Amiot's remark would be repeated by one of the leading anthropologists of the day, Paul Topinard, where it was subtly altered into a claim that the Chinese, much like the West, divided the world into five *races* according to skin color. The fit was not exactly perfect, however, since one of the colors identified by Amiot, the *couleur de chair* (flesh-colored or the color of meat), created a problem for Topinard since it existed alongside white, which was the real "flesh" color as far as the racist West was concerned. In the introduction we have seen a similar pattern in nineteenth-century readings of the "races" depicted in the tomb paintings of the Egyptian pharaoh Seti I, which Topinard also mentioned in his next paragraph as another proof of what he called "the first classifications of the human races."[17]

According to Chinese formulations associated with the Five Elements, moreover, yellow was central, but the other colors hardly fit Western ideas about racial difference. The West was white, but not as in "Caucasian"; white was a withering color, the color of death. The north was black, the reverse of Western ideas about darker skins caused by warmer climates. South was red and the east was green. In a seventeenth-century Japanese map of the world brought to Europe by Engelbert Kaempfer, to give one further example, Asia was labeled as yellow, as might have pleased the West, but Europe was green, Africa was red, and America brownish.[18]

But these concepts were quite different from Western race categorizations; Chinese never referred to themselves as "yellow people," modern phrases like "Sons of the Yellow Emperor" notwithstanding. Older Chinese texts did occasionally describe the colors of different peoples of the world, as in the thirteenth-century catalogue of foreigners included in the *Zhufan zhi,* or *Description of Barbarous Peoples*, by Zhao Rugua. People of the Far West, in this case the Middle East, were characterized as "white," while others in Ceylon, Java, and Malabar were black, black and red, or purple, respectively.[19] Much as in the pre-eighteenth-century European tradition, human lightness and darkness were markers of perceived levels of civilization and not race, with China—the Middle Kingdom, the center of the world—having the whitest people of all. Yellow in China can also carry negative connotations, as in the phrase "yellow skinned and scrawny," which is indicative of decrepitude or disease, and a "yellow movie" or a "yellow joke" (a modern phrase only, probably influenced by the term "yellow journalism" in the West) is vulgar, lascivious, or pornographic.[20]

Yet in the case of describing people of different nationalities in Chinese, it was far more common that foreign people were singled out for their hairiness or their big noses or their deep-set eyes, not their skin color. In 1542, to take just one example, at the beginning of the early modern period of Sino-Western contact, the Portuguese were not allowed to dock in Canton because an edict had been issued barring "the men with beards and big eyes" (*os homens das barbas, e olhos grandes*) from entering the country.[21]

THE RECEPTION OF YELLOW IN JAPAN

In late nineteenth- and early twentieth-century Japan, on the other hand, the notion of a "yellow race" was understood very differently, because in the first place yellow did not have the same cultural or historical connotations. Encyclopedic dictionaries of Mandarin Chinese can fill more than thirty pages describing the many nuances of yellow; in Japanese reference works the same entry covers less than half a page. There even seems to be some doubt as to whether yellow was considered an independent color category in classical Japanese, as for example in the lack of distinction between phrases like "yellow earth" and "red earth."[22]

But more importantly, the notion of the Japanese as "yellow" was received with much more resentment in Japan, since it meant that the country was being lumped together by the West with China, from which many Japanese were also trying to distinguish themselves. Japan, like China, had been keen on importing a great variety of Western scientific thinking since the middle of the nineteenth century, including theories of race and evolution, but while some seemed willing to accept that they were part of the "yellow race"—although far superior to the "yellow Chinese"—others rejected the notion entirely and argued that they had much more in common with the white West than with their neighbors in Asia. One example was the well-known Meiji historian and economist Ukichi Taguchi, who wrote in 1904 that "we can reject the ill repute that the Japanese are yellow"; "those who maintain superiority and excellence in Japanese society are by no means of the yellow race."[23]

The Yellow Peril, it was often argued, was little more than an invention of the Western powers who were jealous of Japan's own imperial aspirations. Its first major acquisition, won at the conclusion of the Sino-Japanese War in 1895, was Taiwan, and in 1902 Japan entered into an official alliance with Great Britain, as if to suggest that it should now be considered on an equal footing with the most powerful "white" nations.[24] As in China some Japanese intellectuals suggested that Russia represented the real "yellow peril," but in Japan this sentiment could go even further to argue that it was also the Russians who were the true "Mongols," a group which, it was proudly asserted, had never conquered Japan.[25]

Discussions of race were also a major theme in foreign policy debates within Japan, which frequently emphasized Japanese identity in opposition to the "white" West, but which at the same insisted that what put Japan on a par with the West was its high level of civilization and culture. Nationalistic Japanese of this period were not likely to emphasize that they were a "yellow" nation, even if they, too, had generally come to accept Western ideas of racializing the world and that their own skin was indeed a yellow color. As a Buddhist representative to a 1904 conference on religion asserted, "Japan has a yellow skin over a white heart," or as the president of Kyoto University remarked in a brief statement included at the end of the entry on Japan for the 1910–11 edition of the *Encyclopaedia Britannica*, the Japanese rejected "discrimination against them as belonging to the 'yellow' race"—even if, he added, "they cannot change the colour of their skin."[26]

In fact many Western commentators seemed to agree that the Japanese were "less yellow" than the Chinese, but only to a point. Questions about the racial makeup of the Japanese have had a long history in the West, and particularly the degree to which they could be considered as descendants of the Chinese (as was frequently claimed in China), since both nations used the same written language and seemed to share many cultural traits. In the early eighteenth century Kaempfer devoted two chapters to this question, concluding that there were many foreign influences in the modern Japanese population. For the nineteenth century the issue was further complicated by the fact that the indigenous, bearded Ainu people seemed to have more in common with the white race than with the yellow Mongolian one.[27] By the second decade of the twentieth century it was not uncommon to find essays in the Western popular press with titles like "Are the Japanese Mongolian?," "Who Are the Japanese?," and "Ethnological Basis of the Japanese Claim to Be a White Race." The Japanese might be lighter or less Mongolian than the Chinese, it was argued, but at the same time they were not quite white either, however Westernized they may have become. At best they could be called "adoptively European." In the apt phrasing of Rotem Kowner, the Japanese during this period were "lighter than yellow, but not enough."[28]

This edition of the *Encyclopaedia Britannica* also bluntly privileged the Japanese over the Chinese, in terms of both skin color and general cultural development. While readers were told that the Chinese were ethnically very mixed and that people from different provinces did not look the same, there was also the by now standard claim that their "complexion varies from an almost pale-yellow to a dark-brown, without any red or ruddy tinge. Yellow, however, predominates." This information was borrowed from a primer on Chinese geography by a Jesuit missionary first published in 1905, although, interestingly enough, the encyclopedia omitted a concluding clause to clarify that "the appellation of '*yellow race*'" applies to "the Japanese, Manchu, and Mongolian races" as well. The *Britannica*'s entry on Japan, on the other hand, was twice as long and omitted references to skin color entirely, offering instead the same tripartite summary of Japanese ethnicity developed by Erwin Baelz that we have mentioned in the previous chapter. An editors' note provided a clear motive for this de-emphasis: Japan deserves "inclusion among the great civilized powers of the world."[29]

A profusely illustrated eighteen-volume world history titled *The Book of History*, which began publication in 1915, was similarly lopsided. A monument to the self-confidence of the European (and especially Anglo-Saxon) imperialistic civilizing mission, the first volume documented among other introductory matters "The Rise of Civilization in Europe" and "The Triumph of Race." An accompanying lexicon called "An Alphabet of Races" identified the Chinese as "one of the most numerous races of the world" but also "the most homogeneous," "yellow-skinned, short in stature"; the Japanese were a subgroup of the "Northern Mongolian family" and were "of short but sturdy stature, white skin, and yellow or sallowish complexion." The differences here were subtle: while Chinese skin was yellow, the Japanese were white with a yellow (or sallow) complexion. While they remained yellow and Mongolian they were also conferred a certain whiteness because, as the author also noted, they were "the most enterprising and civilized people in Asia, often called 'the English of the Far East.'" Germans preferred a different cliché: Prussians of the Far East. The Chinese, however, "possess an ancient and highly organized civilization, which is characterized by its conservatism and slowness to accept new ideas—so different... from the Japanese." An accompanying illustration showed a distinctly lighter-skinned Japanese head when compared to its East Asian neighbors, the Ainu included.[30]

In terms of ethnicity the category known as "the Chinese" is certainly just as thorny, but in the period of the yellow peril, everyone, including many Chinese themselves, and especially those who had studied in the West, seemed willing to accept that they were a "yellow" people. Li Chi's classic 1928 study, *The Formation of the Chinese People*, expressed dissatisfaction that the skin tone of his research subjects did not match the range of yellows included in the scale produced by Felix von Luschan for the purpose of measuring skin color. Li's study also accepted "Sons of the Yellow Emperor" as a legitimate anthropological category. A slightly later essay in the *Zeitschrift für Rassenkunde* by Chungshee H. Liu agreed that Chinese skin varied from "yellowish to yellowish brown with [a] yellow tinge under the skin," although, he added, sometimes the skin was "so dark that the yellow tinge is almost obscured." In other words, if some Chinese people were *not* ostensibly yellow it was simply because this aspect of their skin tone was hidden.[31]

For some Western onlookers Japan was a "yellow hope" rather than a yellow peril, as it was thought that the country might be able to bring needed social and economic change to the entire region, and to China in particular. In 1904 Sun Yat-sen referred to the prospect of a modernized China as a "yellow blessing."[32] But the idea that Japan might actually be able to fill such a role was seen as even more worrisome, especially should it manage to militarize China, thus creating a Pan-Asian alliance and bringing to life an old stereotype: that the Mongol invasions of old had failed because, as Lawrence put it, "though powerful in conquest and desolation, they knew not how to possess and govern." When Kaiser Wilhelm's ambassador presented a version of the Yellow Peril drawing to the tsar in 1895, it was explained that the military reorganization of China's "immense mass" would result in "a struggle for existence between the white and the yellow races." In 1905 one Japanese scholar countered these fears by arguing that "the Japanese are teaching the Chinese nationalism, how to defend their empire and to win the respect of other nations, in opposition to the universal anarchy of the Mongol hordes." "But if the Chinese should pursue a course offensive to western civilization," he continued, "the Japanese would cooperate with Christendom hand in hand against them, just as they did in 1900 when the Boxers treacherously offended against civilization and international law."[33]

Yet for much of "Christendom" both China and Japan were yellow perils, if not always at precisely the same moment or for the same reasons. As a prejudice this found its most insidious manifestation in a series of immigration restrictions enacted in the United States (among other places) against the further influx of "yellow Mongolians," beginning with the Chinese Exclusion Act of 1882 and frequently extended and enlarged thereafter. Despite effusive praise in such texts as the *Encyclopaedia Britannica*, the Japanese were similarly restricted; a 1922 Supreme Court case had ruled that a Japanese immigrant, who had already been living in the United States for twenty years, was ineligible for citizenship since he was not white. In that same year the eminent anthropologist and Smithsonian curator Ales Hrdlicka, himself a naturalized citizen, was called to Congress to testify on the permanently "unassimilable" nature of the Japanese, who, he repeatedly and confidently asserted, were a "yellow-brown or mongoloid people"—a fact with which, he said, all Japanese scientists would concur. Like all East Asians they were seen as fundamentally different both physically and mentally, and the possibility of assimilation by intermarriage was brusquely dismissed as a "deterioration of the blood of the American people."[34]

THE PERSISTENCE OF YELLOW

By the 1920s, as we have seen on a number of occasions before, the yellowness and the perilousness of East Asians had become fully accepted by the West, but once the racialized color terms had become absorbed by Chinese and Japanese writers as well, although in different ways and with different results, it only helped to strengthen the notion even further. Whether it was a positive or negative term, the color of royalty or a yellow peril, a quantifiable skin color or just a cultural symbol, it was subject to considerable variation. I have given only a small sample here. Yet its taken-for-granted nature was traceable even in such mundane and humble objects as the yellow pencil, which at the end of the nineteenth century became firmly associated with the Siberian graphite out of which it was fashioned and was given exotic eastern names such as Koh-I-Noor, Mongol, and Mikado.[35]

Yellowness had been fully naturalized and was perhaps even explainable from an adaptational point of view, as in John L. Myres's introductory contribution for *The Cambridge Ancient History* of 1923, which observed that "the yellow skin-colour of [prehistoric] Mongoloid man gives him protective camouflage in sandy desert and dry-grass steppe." James Cowles Prichard had made much the same claim in the 1840s, and a physician working in China twenty years later analogously hypothesized that bad smells and dust storms had produced the small noses and slitted eyes of the local population.[36] In 1911 one sociologist remarked how the loess deposits of northern China covered the entire region with a yellow dust, including "the yellow people and their clothing," and in 1924 English anthropologist Alfred Haddon took Prichard's terms for hair color and transferred them to the color of human skin instead. Although Haddon said he was hesitant about skin color classifications generally, yellow-skinned *xanthodermi*, he argued, were "of varying tints from quite light to brown, but always with a yellowish tinge."[37]

The idea began to be widely disseminated in popular introductions to anthropology. Haddon's was one such work. Another was by E. P. Stibbe, published in 1930, which remarked that any student "without technical knowledge" and "based on his own observations of those around him" would certainly be able to perceive that there are white, yellow, and black people. While white and black were regularly put in quotation marks, since white people weren't really white and blacks not black, the "Mongoloid race, as exemplified by a Chinaman," was yellow without any sort of qualification. Harvard anthropologist Earnest Hooton was hesitant about how to classify what he called "the inscrutable Mongoloids," but he, too, never doubted the reality of their "yellowish skin color." At Columbia, Ralph Linton lamented that the Mongolian race had been "used as a catch-all for races and stocks which clearly were not Negroid but which the Caucasian scholars were unwilling to admit to their own select company." But, he added, "the North Chinese race is tall, round-headed, with light yellow skins, small, straight noses, thin lips, and slant eyes." And Roland Dixon, also of Harvard, agreed that while "the population of the whole of China is light," it must be qualified as having "a characteristic yellowish tinge." Japanese skin, somewhat higher on the color scale, was "usually fair and slightly yellowish." Both populations could also embody "rosy" complexions, he continued, but if in some parts of China people "have a white-rosy complexion quite like that of a European," these were merely "border tribes." Old stereotypes die hard: the notion that the only "well-proportioned" Chinese were those with non-Chinese ancestors can be found in the very first European book on the empire by Gaspar da Cruz in 1569. Even worse, Dixon concluded that the yellowishness of the Chinese "early won for them the name of the Yellow Race," even though, as I have tried to show in the chapters of this study, that particular notion can be traced back only (at best) to the very end of the eighteenth century.[38]

It is important to note that these citations represent only a small selection of a far larger corpus of works that could easily be extended and is hardly confined to English-language publications. Moreover, these are statements drawn not from fringe or reactionary authors but from some of the leading and most influential academics of the day. Many other examples have been cited in the previous chapters, when taxonomers, comparative anatomists, anthropologists, and physicians decided that the Mongolian race was very much yellow indeed. The question is not whether "yellow" or "yellowish" or "yellowish tinge" is an accurate descriptor, but rather the values that these or any other color terms are regularly invested with. According to some in East Asia yellow was a sign of superiority or the glories of an ancient and unbroken civilization. But in the West it was part of a larger peril embodied by anyone who was not white. Yellow was simply one more term in a quantitative hierarchy of "civilization" that seemed to require no further explanation, and if it turned out that yellow was a concept that the Chinese, at least, also embraced, so much the better.

In 1910 Franklin Giddings, a distinguished professor of sociology at Columbia, published what he termed a "scientific scale" of the racial makeup of the United States, purporting to show how far different racial groups were from "native born of native parents" (Anglo-Saxon is implicit). The results showed that "Asian yellow" people came in seventh place out of nine, with only "civilized dark" and "uncivilized" being lower (figure 16). As was generally the case since the early nineteenth century, of all the nonwhite races the yellow one was awarded the highest status. Yet looking closely at Giddings's criteria we see that the only thing that really distinguished East Asians from "all other whites" (specified as Turks, Persians, and African whites), was that white people possessed a color "lighter than yellow."[39]

Even after World War II, when the world was supposed to be entering a new era of racial tolerance, UNESCO's official statement "The Race Question" (1950)—produced by an international team of anthropologists and designed to demystify race and its attendant hierarchies—still identified "Mongoloid," "Negroid," and "Caucasoid" as the three "major divisions" that "most anthropologists agree on." In the following year one of the statement's main authors, Ashley Montagu, best known for his debunking of race in *Man's Most Dangerous Myth*, published an extended commentary in which he added

MARKING-SCHEME FOR NATIONALITIES

	Parents Native-born	Self Native-born	Native Language English	Reared under Celto-Teutonic Traditions and Culture	Reared under Constitutional Government	Of European Stock and Reared under European Civilization	Belongs to Race that Has Created an Independent Political State	Belongs to Race that Has Created an Ethical Religion	Belongs to Race that Has Created a Literature	Belongs to Race that Has Independently Risen above Barbarism	Color Lighter than Yellow	Color Lighter than Red	Color Lighter than Brown	Color Lighter than Black	Total Mark	Position
Native-born of native parents	1	1	1	1	1	1	1	1	1	1	1	1	1	1	14	0
Native-born of foreign parents	..	1	1	1	1	1	1	1	1	1	1	1	1	1	13	1
Foreign-born, English speaking	..	..	1	1	1	1	1	1	1	1	1	1	1	1	12	2
Northwestern European	..	..	..	1	1	1	1	1	1	1	1	1	1	1	11	3
Southern European	..	..	..	..	1	1	1	1	1	1	1	1	1	1	10	4
Eastern European	..	..	..	..	..	1	1	1	1	1	1	1	1	1	9	5
All other whites	..	..	..	..	..	..	1	1	1	1	1	1	1	1	8	6
Asian yellow	..	..	..	..	..	..	1	1	1	1	..	1	1	1	7	7
Civilized dark	1	1	1	1	1	..	..	..	..	..	..	..	..	..	5	8
Uncivilized	..	..	..	..	..	..	..	..	..	..	..	..	1	1	2	9

THE SOCIAL MARKING SYSTEM

Figure 16: "Marking-Scheme for Nationalities," from Franklin H. Giddings, "The Social Marking System" (1910). The top of the table lists features of civilized cultures (according to Giddings) and then judges whether each of the racial groups on the left has achieved them. At the right the results are tabulated, with "native born of native [Anglo-Saxon] parents" receiving a perfect score. "Asian yellow" people fare better than "Civilized dark" and "Uncivilized," because of their high level of politics, religion, and literature, but "All other whites" score even higher because their "Color [is] Lighter than Yellow." Princeton University Library.

that Mongoloids were characterized by sparse black hair, epicanthic folds, and skin with "a faintly yellowish tinge." A small booklet published by UNESCO in 1952 titled *What is Race?* agreed that there was "a group of yellow-skinned people in North-East Asia," and that "Mongoloid" skin color was "pale yellow to yellow-brown," although "reddish brown" could also be seen.[40] While these texts finally began to call into question long-standing racial prejudices that had pervaded the West for centuries, it is telling that both yellow and Mongoloid remained almost competely unexamined categories.

Or to take one final example, in a long essay on physical anthropology included in the *Encyclopédie de la Pléiade* volume on general ethnology in 1968, Henri Victor Vallois still insisted on white, yellow, and black divisions for mankind—even as, he added, the terms "white" and "black" were only approximations. And yellow? When one hundred pages later he came around to a discussion of "the great yellow race" he admitted that yellow skin was an even less tidy category than white or black, since it ranged from light yellow to a dark brown in which yellow was "completely lacking." What bound these people together, however, were "other dispositions" that included black hair, a particular shape of the body and face, and so on. One might well wonder: why not drop the term "yellow race" entirely? Perhaps because, as Vallois continued, in the "Central Mongolian Race," which includes both Chinese and Central Asians, "the yellow characteristics are well marked, and it is also here that a yellow skin tone is the most pronounced: a wheaten yellow" (*d'une jaune froment*). In terms of color prejudice we hardly seem to have progressed beyond Blumenbach's characterization of the Mongolian race as "halfway between grains of wheat and cooked quinces" in 1795.[41] Perhaps Vallois, one of the grand old men of French anthropology and a chief editor of the *Revue d'anthropologie* since 1932, was already out of date, but there can be little question that the notion of yellow, to say nothing of the Mongolian, had become so entrenched that it was difficult to leave behind.

Yet we have seen in the chapters of this book that yellow and Mongolian/oid were both recent and somewhat mysterious choices. Once these fantasies had been understood as defining a racial category, and once they had become allied to sometimes very different notions of yellow (and Mongolian) within East Asian contexts, the terms took on an air of certainty and consensus that was not easy to dislodge. I am unable to speak for any author who claims to see yellow when he or she looks at East Asian skin (I, for one, do not), and yet it seems to me much more important to understand that this color term, like Mongolian and its variants, is hardly neutral and that it has had a complex and often surprising history. This can easily lead to inaccurate statements about its complicated and often fitful development, especially when it is assumed to be far older than it actually is, as in the following two very recent examples:

> The "yellow peril" has been part of a Western image of China ever since Genghis Khan's invasion of Europe.
>
> The notion of "yellow peril" was first used in Europe in response to the sixth-century invasion of Attila the Hun.[42]

The invasions of Genghis Khan or Attila the Hun were certainly perceived as perils in terms of the West's image of itself as the bearer of the true religion and the apex of human civilization (although

"Europe" was a somewhat later development). But these perils were not considered yellow until the nineteenth century, after Mongolian, that bizarre and inappropriate means of aggregating the varied peoples and cultures of the East Asian region, had become a racial category. And after an astonishing coincidence with yellow in China (but only China), color and race and perilousness came together to express a particularly virulent form of racism in the late nineteenth century. The real question, which is also the question expressed in the title of this book, is how and why that peril had become coded as yellow in the first place.

Notes

INTRODUCTION
No Longer White: The Nineteenth-Century Invention of Yellowness

1. Isidore, *Etymologiae*, 82:497; Beazley, *The Dawn of Modern Geography*, 2:549–642. See Miller, *Mappaemundi*, 1:53, for an illustration of the "St. Sever" map (ca. 1030), which also identified India as having people *tincti coloris*.

2. Dante, *La divina commedia*, ed. Lombardi, 1:481.

3. Blumenbach, *De generis humani varietate nativa*, 3rd ed., 120 (*The Anthropological Treatises of Johann Friedrich Blumenbach*, 209); Goldsmith, *An History of the Earth*, 2:232. This view became commonplace, as in Humboldt, *Voyage aux régions équinoxiales du nouveau continent*, 3:286; it was later criticized by Darwin in *Expression of the Emotions in Man and Animals*, 315–20.

4. Lanci, *Paralipomeni all'illustrazione della sagra scrittura*, 2:273; Tylor, *Anthropology*, 68. A red Adam was also mentioned in Labat, *Nouvelle relation de l'afrique occidentale*, 2:256.

5. Dante, *La divina commedia*, ed. Lombardi, 1:481.

6. Dante, *The Comedy of Dante Alighieri*, trans. Sayers, 1:290. On "the curse of Ham" see Braude, "The Sons of Noah and the Construction of Ethnic and Geographical Identities in the Medieval and Early Modern Periods."

7. Dante, *Divine Comedy*, trans. Musa, 2:448–49; Dante, *Inferno*, trans. Zappulla, 307. The most authoritative modern interpretations, which reject readings based on skin color, are Singleton, "Commedia, Elements of Structure"; and Freccero, "The Sign of Satan." See also Graf, *Miti, leggende e superstitioni del medio evo*, 2:92–93; and Toynbee, *A Dictionary of Proper Names and Notable Matters in the Works of Dante*, 402.

8. Basic introductions to the tomb and its environs include Hornung, *The Valley of the Kings*; and Weeks, *The Valley of the Kings*. The most detailed study is Hornung, *The Tomb of Pharaoh Seti I*.

9. Belzoni, *Narrative of the Operations and Recent Discoveries within the Pyramids*, 242–43, 528. On the exhibition see Pearce, "Giovanni Battista Belzoni's Exhibition of the Reconstructed Tomb of Pharoah Seti I"; and on Belzoni's showmanship see Mayes, *The Great Belzoni*.

10. Champollion, *Lettres écrites d'Égypte*, 248, 250.

11. Hornung, *Ägyptische Unterweltsbücher*, 233–35; Hornung, *The Ancient Egyptian Books of the Afterlife*, 62.

12. Champollion, *Lettres écrites d'Égypte*, 249–51.
13. Blumenbach, "Observations on Some Egyptian Mummies," 193; Winckelmann, *The History of Ancient Art*, 1:167–68.
14. Champollion-Figeac, *Égypte ancienne*, 29.
15. Camper, *Works*; Virey, *Histoire naturelle du genre humain*; Bory de Saint-Vincent, *L'homme*; Morton, *Crania Aegyptiaca*; Brugsch, *Histoire d'Égypte*, esp. 1–4.
16. Minutoli, *Reise zum Tempel des Jupiter Ammon*, plate 3; Rosellini, *I monumenti dell'Egitto e della Nubia*, plates vol. 1, plates 155–56; Lepsius, *Denkmaeler aus Aegypten und Aethiopien*, vol. 6, plate 136. Rosellini had rejected Champollion's conjecture that the "Tamhou" were primitive Europeans, insisting instead that they represented two groups of Asians, one more Eastern than the other. He also noted that one was yellow and one was white and emphasized this in another plate (vol. 1, plate 110) comparing heads of different "foreigners" shown in the tombs: see *Monumenti*, part 1, vol. 4, pp. 228–40. Lepsius described "Namou" figures from another room in the same tomb as Asians with *gelbbrauner* skin color: *Denkmäler aus Aegypten, Text*, 3:218. In Rosellini an (unpaged) index of plates identified the figures as "Le diverse specie di uomini noti agli Egizi." Regarding Champollion's *Monuments de l'Égypte et de la Nubie*, vol. 3, plates 238–40, an accompanying explanation identified the plates as "an ethnographic tableau of the four parts of the world according to the Egyptians" (3:2); the "Europeans" had *peau blanche*.
17. Lefébure, "Les races connues des Égyptiens," 66.
18. Nott and Gliddon, *Types of Mankind*, 85, 720 n. 179.
19. Poole, "The Egyptian Classification of the Races of Man," 376; Meyer, "Bericht über eine Expedition nach Ägypten zur Erforschung der Darstellungen der Fremdvölker"; Wreszinski, *Atlas zur altägyptischen Kulturgeschichte*; Topinard, *L'anthropologie*, 364–65 (*Anthropology*, 344); Topinard, *Éléments d'anthropologie générale*, 63.
20. Nott and Gliddon, *Types of Mankind*, 86; Winchell, *Preadamites*, 209; Haddon, *The Races of Man*, 2 (cf. Haddon, *The Study of Man*, 16); Rogers, *The Colors of Mankind*, 22. See also Young, *Colonial Desire*, 76–78.

CHAPTER 1
Before They Were Yellow: East Asians in Early Travel and Missionary Reports

1. Hodgen, *Early Anthropology in the Sixteenth and Seventeenth Centuries*; Devisse and Mollat, *The Image of the Black in Western Art*; Hahn, "The Difference the Middle Ages Makes"; Bartlett, "Medieval and Modern Concepts of Race and Ethnicity."

2. Snowden, *Before Color Prejudice*; Hall, *Ethnic Identity in Greek Antiquity*; Isaac, *The Invention of Racism in Classical Antiquity*; Wittkower, *Allegory and the Migration of Symbols*, 45–92; Friedman, *The Monstrous Races in Medieval Art and Thought*.

3. Ramusio, *Delle navigationi et viaggi*, 2:21, 46, 50.

4. Odoric's account was published in Ramusio's third edition of 1583; I have used the modern edition of Ramusio instead: *Navigazioni e viaggi*, 4:284.

5. Isidore, *Etymologiae*, 82:497. A recent English translation renders the phrase as "people of color," a misleading piece of modern jargon (*Etymologies*, 286). One of Isidore's main sources was Solinus, writing in the third century, who specified that people who lived in the East were "burned by the heat to a degree beyond other people," and that their color "bespoke the force of the climate" (*Collectanea rerum memorabilium*, 186).

6. Zarncke, "Der Priester Johannes"; Ross, "Prester John and the Empire of Ethiopia"; Letts, "Prester John"; Slessarev, *Prester John*; Rachewiltz, *Prester John and Europe's Discovery of East Asia*.

7. Montalboddo, *Paesi novamente retrovati*, sig. H4. For an Italian text in manuscript see Radulet, *Vasco da Gama*, 174. On the Portuguese arrival in the region see Cordier, "L'arrivée des Portugais en Chine"; Kammerer, "La découverte de la Chine par les Portugais"; Schurhammer, "O descobrimento do Japão pelos Portugueses"; and Loureiro, *Fidalgos, Missionários e Mandarins*.

8. Needham, *Science and Civilisation in China*, 4:3:508.

9. Camões, *The Lusiads*, 1:277–79, with the Portuguese original on facing pages.

10. Cited in Lach and Van Kley, *Asia in the Making of Europe*, 1:731; for the Portuguese text see Sena, "Macau," 34.

11. Pires, in Ramusio, *Delle navigationi et viaggi*, 1:372v; for the Portuguese text see Pires, *Suma Oriental*, 2:393. See also letters from Andrea Corsali (1515) and Maximilianus Transylvanus (1522), also published in Ramusio (*Delle navigationi et viaggi*, 1:197v [misnumbered as 198v], 1:384). The former noted that the Chinese were "of our quality," while the latter stressed their whiteness as well. Maximilianus's report had already been published in *De Moluccis insulis* (1523) and then again in *Novus orbis regionum ac insularum veteribus incognitarum* (1532). Pigafetta, who participated in Magellan's circumnavigation in 1519–22, also observed that the people of China were white (*Le voyage et navigation faict par les Espaignols es Isles de Mollucques*, sig. 73v).

12. Ramusio, *Delle navigationi et viaggi*, 1:354v; for the Portuguese text see Barbosa, *Livro em que dá relação do que viu e ouviu no Oriente*, 217. The Japanese language was also compared to German in Escalante Alvarado, "Relación del viaje que hizo desde la Neuva-España á las islas del Poniente,"

203; see also Dahlgren, "A Contribution to the History of the Discovery of Japan," 245.

13. Graberg da Hemsö, "Lettera di Giovanni da Empoli," 59; Pires, *Suma Oriental*, 2:461; Lopes de Castanheda, *História do descobrimento & conquista da India pelos Portugueses*, 2:447. Castanheda similarly noted that the Japanese of Lequia were "white [*alva*] and beautiful and very well dressed." See also Corrêa's manuscript from about 1550, published as *Lendas da India*, 2:529, which described the Japanese as white (*branca*); as well as an unpublished report from 1518 cited in Schurhammer, "O descobrimento do Japão," 514. In 1548 a Jesuit account noted that the Japanese were "white like us" (*bianchi como noi*): *Documentos del Japón*, 62.

14. Couto, *Da Asia*, 13:265–66. See also Ferguson, "Letters from Portuguese Captives in Canton," 437. Other versions of the story were told in Galvão, *The Discoveries of the World*, 229–30 (with the Portuguese original); and Lucena, *História da vida do padre Francisco de Xavier*, 461–62. For Alvares see Pires, "O Japão no seculo XVI," 57; Xavier, *Epistolae*, 2:277.

15. Xavier, *Epistolae*, 2:186; Gago, in *Cartas . . . de Japão & China*, 1:40 (and in Spanish in *Cartas . . . de Japon*, sig. 71v); Hay, *De rebus japonicis*, 7. On Xavier's influence see Schurhammer, "Der 'Grosse Brief' des Heiligen Franz Xaver." A few other period examples include letters by Gonzalo Fernández, Luis Fróis, and Gaspar Vilela, all in *Cartas . . . de Japão & China*, 1:73v, 1:172, and 1:193v; Torsellino, *De vita S. Francisci Xaverii*, 187; and Lucena, *Francisco de Xavier*, 469. The detail also began to appear in more general compilations such as Botero, *Relationi universali*, 2:10, although oddly enough it was removed from the English versions: *The Worlde*, 198; and *Relations of the Most Famous Kingdomes and Common-wealths Thorowout the World*, 621.

16. Valignano, *Sumario de la cosas de Japón*, 200; van Noort, *De reis om de wereld*, 1:115; Ijzerman, *Dirck Gerritsz Pomp*, 21.

17. *Relatione del viaggio et arrivo in Europa et Roma de' principi giapponesi*, sig. B3; Berchet, "Le antiche ambasciate giapponesi in Italia," 151, 174; Benacci, *Avisi venuti novamente da Roma delli XXIII de marzo 1585*; Boscaro, "The First Japanese Ambassadors to Europe," 10 n. 37; Gualtieri, *Relationi della venuta degli ambasciatori giaponesi à Roma*, 157. See also Boscaro, *Sixteenth-Century European Printed Works on the First Japanese Mission to Europe*; Brown, "Courtiers and Christians"; and Cooper, *The Japanese Mission to Europe*. A Japanese known as Bernardo also visited Rome for nearly a year in 1555, but little is known about his stay there: see D'Elia, "Bernardo, il primo giapponese venuto a Roma."

18. Berchet, "Le antiche ambasciate giapponesi," 190, 193; Nakamura, "Passage en France de Hasekura," 452, 454. On this embassy see Meriwether, "A Sketch of the Life of Date Masamune."

19. Bartoli, *L'Asia*, 3:7; Charlevoix, *Histoire de l'établissement, des progrès et de la décadence du Christianisme dans l'empire du Japon*, 1:5; Charlevoix, *Histoire et description générale du Japon*, 1:54. For material on Japan generally, an excellent resource is *Japan in Europa*.

20. Kaempfer, *Geschichte und Beschreibung von Japan*, 1:110; Kaempfer, *The History of Japan*, 1:95; Kaempfer, *Histoire . . . du Japon*, 1:83. Kaempfer's Dutch translator chose *galachtig* (i.e., yellowish) instead: *De beschryving van Japan*, 68. For Kaempfer's influence in the eighteenth century see Prévost, *Histoire général des voyages*, 10:576; Walbaum, *Ausführliche und merkwürdige Historie der Ost-Indischen Insel Groß-Java*, 202–3; and Marsy, *Histoire moderne des Chinois . . .*, 2:404–5. Interestingly, the German version of Prévost translated *olivâtre* as *bleyfärbig* (i.e., lead-colored): *Allgemeine Historie der Reisen zu Wasser und Lande*, 11:595. For the prevalence of "tawny" in English cf. Benjamin Franklin, "Observations Concerning the Increase of Mankind," first printed in 1755 (*Papers*, 4:234): "the Number of purely white People in the World is proportionably very small. All Africa is black or tawny. Asia chiefly tawny. America (exclusive of the new Comers) wholly so."

21. Xavier, *Epistolae*, 2:277, 291; Hay, *De rebus japonicis*, 883, 895; Purchas, *Purchas His Pilgrimes*, 3:320; Torsellino, *De vita S. Francisci Xaverii*, 241. See also a 1546 letter in *Documentação para a história das missões do padroado Português do Oriente*, 3:379; an anonymous 1548 manuscript in *Livro que trata das cousas da India e do Japão*, 116; an unpublished manuscript by Loarca, "Relación del viaje que hezimos a la China desde la ciudad de Manila en 1575," ch. 3; and Guzmán, *Historia de las misiones de la Compañía de Jesús en la India Oriental*, book 3, ch. 40, p. 156. Other influential published sources stressed both Chinese whiteness and their "heathen" nature, such as Münster, *Cosmographia*, 1095 (in English as *A Treatyse of the Newe India*, sig. F2v); and Eden, *The History of Travayle in the West and East Indies*, sig. 249v (also in Hakluyt, *Principal Navigations*, 2:2:78). See also Boxer, *South China in the Sixteenth Century*, 38.

22. Ricci, *De Christiana expeditione apud Sinas*, 85–86. An abbreviated version in English appeared in Purchas, *Purchas His Pilgrimes*, 3:394; and a full version in 1953 as Ricci, *China in the Sixteenth Century*. For Ricci's original manuscript text see *Fonti Ricciane*, 1:88. The earliest published references to Chinese skin color variation seem to be Osório, *De rebus Emmanuelis*, 3:343 (in English as *The History of the Portuguese during the Reign of Emmanuel*, 2:246); and Escalante, *Discurso de la navegacion . . . à los Reinos y Provincias del Oriente*, sig. 42–43 (*A Discourse of the Navigation . . . to the Realmes and Provinces of the East Partes of the Worlde*, sig. 21–21v). On the Jesuit policy of accommodation in China see Young, *Confucianism and Christianity*; Rule, *K'ung-tzu or Confucius?*; Mungello, *Curious Land*; and Standaert, *The Interweaving of Rituals*.

23. Ricci's lament is cited in Boxer, *The Christian Century in Japan*, 78. On Chinese children see Rada, *Relaçion verdadera delascosas del reyno de Taibin por otro nombre China*, fol. 25 (Boxer, *South China*, 282); Lucena, *Francisco de Xavier*, 861; Du Jarric, *L'histoire des choses plus mémorables advenuës tant és Indes Orientales*, 732–33; Bartoli, *La Cina*, 67; and Du Halde, *Description... de l'empire de la Chine*, 2:80. After Du Halde it became a commonplace detail, as in *A New Collection of Voyages and Travels*, 4:73. The subject is also mentioned in a manuscript letter by one of Ricci's contemporaries in China (cited in Bartoli, *La Cina*, 67 n. 61).

24. Martini, *Novus atlas sinensis*, 7; Sánchez, "De la entrada de la China en particular," 1:443 ("The Proposed Entry into China, in Detail," 6:219). Sánchez's scheme was effectively laid to rest the following year when the Spanish armada was defeated off the English coast. See also Doyle, "Two Sixteenth-Century Jesuits and a Plan to Conquer China." In 1615 de Feynes also referred to both the Cantonese and the South Vietnamese as white, as did Gemelli Careri at the other end of the century: Feynes, *An Exact and Curious Survey of all the East Indies*, 29–30 (and also appearing in Purchas, *Purchas His Pilgrimes*, 3:410); Gemelli Careri, *Giro del mondo*, 4:364. For a manuscript text of de Feynes see *Le voyage de Montferran de Paris a la Chine*, 29–31.

25. Brown and swarthy: Mendoza, *The Historie of the Great and Mightie Kingdome of China*, 4, 19; swarthy: Heylyn, *Cosmographie*, 865; red and tawny: Escalante, *Discourse of the Navigation*, sig. 21; brown and black: Linschoten, *Discours of Voyages into ye Easte & West Indies*, 1:40; tawny and dun: Le Comte, *Memoirs and Observations*, 127; black and tanned: Palafox y Mendoza, *The History of the Conquest of China by the Tartars*, 547; ashy: Dampier, *A New Voyage Round the World*, 1:407; black and sanguine: d'Avity, *The Estates, Empires & Principallities of the World*, 719; olive and brown and ruddy: Du Halde, *A Description of the Empire of China*, 1:281; florid: Du Halde, *The General History of China*, 2:138; deep brown and brunette: Barrow, *Travels in China*, 184.

26. A few examples not mentioned elsewhere in these notes include Ortelius, *Theatrum orbis terrarum*, fol. 93; Nieuhof, *Het gezantschap... aan den grooten Tartarischen Cham*, part 2, p. 56; Dapper, *Gedenkwaerdig bedryf der Nederlandsche Oost-Indische maetschappye... in het keizerrijk van Taising of Sina*, part 2, p. 251; Andersen, *Orientalische Reise-Beschreibungen*, 135; and Semedo, *Relatione della grande monarchia della Cina*, 31. For the eighteenth century this information was canonized by Le Comte, *Nouveaux mémoires sur l'état présent de la Chine*, 1:265–67; and by Du Halde, *Description de la Chine*, 2:80.

27. Mendoza, *Historia... del gran reyno de la China*, 5.

28. Mendoza, *Historia... del gran reyno de la China*, 21–22.

29. Mendoza, *L'historia del gran regno della China*, 25–26, 45; Mendoza, *De historie ofte beschrijvinghe van het groote rijck van China*, 22, 42;

Mendoza, *Histoire du grand royaume de la Chine*, sig. 3v, 15v; Mendoza, *The Historie of the Great and Mightie Kingdome of China*, 4, 19.

30. For other period translations of *rubio* see Howell, *Lexicon tetraglotton*; and Stevens, *A New Spanish and English Dictionary*. In Spanish as in Latin there seemed to be considerable overlap in words for blonde and red, particularly with respect to hair color. See André, *Étude sur les termes de couleur dans la langue latine*, esp. 128–30.

31. Mendoza, *Ein neuwe . . . Beschreibung dess . . . Königreichs China*, 6, 27; Mendoza, *Nova et succincta . . . historia de . . . regno China*, 32, 60.

32. On Native Americans see Vaughn, *Roots of American Racism*, 3–33; Shoemaker, "How Indians Got to Be Red"; Shoemaker, *A Strange Likeness*; and Kupperman, *Indians and English*. On their supposed similarity to the Chinese, especially in terms of color, see García, *Origen de los Indios*, 239–48; Solórzano Pereira, *Política indiana*, 16–22; and Huddleston, *Origins of the American Indians*. On other regions of the world see Bonnett, "Who Was White?"

33. López de Gómara, *Historia general de las Indias*, sig. 96; Eden, *Decades of the Newe Worlde*, sig. 310v; Eden, *History of Travayle*, sig. 4–4v; Eden, *De novo orbe*, sig. 4–4v. López de Gómara's French translator agreed that *tiriciado* was olive (*Histoire generalle des Indes Occidentales*, sig. 250).

34. The only such references I'm aware of are rather late. One is by Raffles, *The History of Java*, 1:341, 344, 351, who noted that some Javanese women enhanced their appearance with the application of yellow powder; all the Javans, he added, "as well as other eastern islanders, may be considered rather as a yellow than a copper-coloured or black race" (1:59) (my thanks to Michael Laffan for this reference). The other is by Finlayson, *The Mission to Siam and Hué*, 227, who remarked that the women of Vietnam had a "yellow complexion" that they "take pleasure in heightening by the use of a bright yellow wash or cosmetic." Perhaps these references were variations on the whitening cosmetics frequently mentioned in sources on China and Japan. See for instance Saris, "Course and Acts to and in . . . Japan," in Purchas, *Purchas His Pilgrimes*, 1:367 (and Saris, *The Voyage of Captain John Saris to Japan, 1613*, 84); Le Comte, *Nouveaux mémoires*, 1:267; and Du Halde, *Description de la Chine*, 2:80.

35. In addition to López de Gómara, quince was also mentioned by Solórzano Pereira, *Política indiana*, 21; and by Bodin, *Les six livres de la république*, 542 (*The Six Bookes of the Common-Weale*, 568). On lemon peel see Blumenbach, *De generis humani varietate nativa*, 3rd ed., 120 (*Anthropological Treatises*, 209).

36. Montanus, *Gedenkwaerdige gesantschappen der Oost-Indische Maatschappy . . . aan de kaisaren van Japan*, 57; Montanus, *Ambassades mémorables de la Compagnie des Indes Orientales . . . vers les Empereurs du Japon*, 53. The English translation used "sallow" instead: *Atlas Japannensis*, 77. Two earlier examples of the Japanese as yellow were Andersen, in Mandelslo,

Morgenländischen Reyse-Beschreibung, 247; and Wernhart, *Christoph Carl Fernberger*, 124–25 (this one was not published). Montanus's description was repeated in Struys, *Drie aanmerkelyke reizen*, part 1, p. 65. It also appeared in Hazart, *Kirchen-Geschichte*, 231, although in the original Dutch text, *Kerckelycke historie van de gheheele wereldt*, the skin color of the Japanese was not mentioned. I am indebted to Peter Kapitza for this reference.

37. Maffei, *L'histoire des Indes Orientales*, 234; Maffei, *Historiarum indicarum*, 144; Maffei, *Le istorie dell'Indie Orientali*, fol. 97; Saris, in Purchas, *Purchas His Pilgrimes*, 1:367 (Saris, *Voyage*, 84); Bacon, *Silva sylvarum*, 2:577; *Allerhand so lehr- als geist-reiche Brief, Schrifften und Reis-Beschreibungen*, vol. 5, part 34, no. 673, p. 37. Valentijn's *Oud en nieuw Oost-Indien*, 2:258, described the Chinese as "not white," "brownish yellow and pale," and "pale yellow" (*niet blank, bruyngeelen en bleek, bleekgeel*). Williams, one of most influential authorities on China for the second half of the nineteenth century, called them "sickly white" (*The Middle Kingdom*, 1:36), a verdict soon repeated with respect to Chinese women in Müller, *Allgemeine Ethnographie*, 363–64.

38. Linnaeus, *Systema naturae*, 10th ed., 1:21.

39. Nichols, *Illustrations of the Literary History of the Eighteenth Century*, 5:318–19, which also provided the extract from the *Gentleman's Magazine*. On other Chinese visitors to England during the period see Appleton, *A Cycle of Cathay*, 130–36.

40. Washington, *Papers*, 3:205, 214; Washington, *Writings*, 28:238–39.

41. Wernhart, *Christoph Carl Fernberger*, 124–25, 137.

42. Berchet, "Le antiche ambasciate giapponesi," 174; Boscaro, "First Japanese Ambassadors," 10 n. 36; Gutierrez, *La prima ambascieria giapponese in Italia*, 67.

43. Sande, *De missione legatorum japonensium*, 403–9. See also Massarella, "Envoys and Illusions." For Valignano on Japanese Jesuits see *Sumario*, 182. The problem of a native clergy was fundamental to Jesuit programs in both Japan and China: Bontinck, *La lutte autour de la liturgie chinoise*; Boxer, "European Missionaries and Chinese Clergy." For Valignano on Japanese whiteness see also his *Historia del principio y progresso de la Compañía de Jesús en las Indias Orientales*, 111, 126–27; as well as Schütte, *Valignano's Mission Principles for Japan*, 1:1:293; Boxer, *Christian Century in Japan*, 84–85; and Moran, *The Japanese and the Jesuits*, 51–53, 97–98, 152, 179.

44. Fryer, *A New Account of East-India*, 155.

45. Pantoja, *Relación de la entrada de algunos padres . . . en la China*, 107 ("A Letter of Father Diego de Pantoja," in Purchas, *Purchas His Pilgrimes*, 3:367).

46. It was another nineteenth-century stereotype that nonwhite people looked more alike than those of the "higher races," as in Hunt, "Race in Legislation and Political Economy," 129. When Trigault's edition of Ricci's journals was translated into English in 1953, the Latin phrase "Sinica gens ferè albi

coloris est" (that is, "Chinese people are as a rule white") was revealingly mistranslated as "Chinese people are almost white" (Ricci, *De Christiana expeditione*, 85; Ricci, *China*, 77). No seventeenth-century translation had made the same mistake (Ricci, *Histoire de l'expedition chrestienne au royaume de la Chine*, 69; Ricci, *Istoria de la China*, fol. 42; Ricci, *Entrata nella China de' padri della Compagnia del Gesù*, 105; Purchas, *Purchas His Pilgrimes*, 3:394).

47. *Encyclopaedia Britannica*, 11th ed., 6:174. This information was quoted from Richard, *Comprehensive Geography of the Chinese Empire*, 340–41, which had originally appeared as *Géographie de l'empire de Chine*, 309–10.

48. *Lettres édifiantes et curieuses*, 26:33; Hume, *Essays*, 275; Grosier, *Description générale de la Chine*, 2:358 (*A General Description of China*, 2:372); Gützlaff, *A Sketch of Chinese History*, 1:43–44. Grosier's text had originally appeared as the final two volumes of Mailla, *Histoire générale de la Chine*; see 13:688.

49. Candidius, "Discours ende cort verhael van 't eylant Formosa," 3:2; Dapper, *Gedenkwaerdig bedryf*, part 1, p. 20; *A Collection of Voyages and Travels*, 1:527; Montanus [i.e., Dapper], *Atlas Chinensis*, 21; Keevak, *The Pretended Asian*, 35–59.

50. Thunberg, *Inträdes-tal, om de mynt-sorter . . . uti Kejsaredömet Japan*, 8; Thunberg, *Resa uti Europa, Africa, Asia*, 3:279; Thunberg, *Reise durch einen Theil von Europa, Afrika und Asien*, 2:154; Thunberg, *Voyage en Afrique et en Asie*, 411; Thunberg, *Travels in Europe, Africa, and Asia*, 3:251; Thunberg, *Voyages . . . au Japon*, 3:193; Siebold, *Nippon*, 1:282; MacFarlane, *Japan*, 141. See also Thunberg, *Verhandeling over de Japansche natie*, 7; and Thunberg, "Ett kort utdrag af en journal, hållen på en resa til och uti Keisaredömet Japan," 147. A secret report on Japan by Raffles from about 1812 (but which remained unpublished until 1929) was exceptional in its description of Japanese color. He reminded his auditors that "the complexion is fair and indeed blooming, the women of the higher classes being equally fair with Europeans and having the bloom of health more generally prevalent among them than is usually found in Europe." The Japanese character, he concluded, "would progressively improve until it attained the same height of civilization with the European": Raffles, *Report on Japan to the Secret Committee of the English East India Company*, iv–v; also reprinted in Paske-Smith, *Western Barbarians in Japan and Formosa in Tokugawa Days*, 131. In this version the Japanese complexion is "perfectly fair." See also Kowner, "Skin as a Metaphor." Wooler's radical newspaper *The Black Dwarf*, which began publication in 1817, featured a number of fictional letters addressed to "the Yellow Bonze of Japan," frequently addressed as "my yellow friend."

51. Moges, *Souvenirs d'une ambassade en Chine et au Japon*, 310; Lawrence, *China and Japan and a Voyage Thither*, 216; Eden, *Japan*, 240; Prichard, *Researches into the Physical History of Man*, 22; Prichard, *The Natural History*

of Man, 230. The darkening of the Japanese in the course of the nineteenth century was also epitomized by the difference between Parker's *Journal of an Expedition from Singapore to Japan*, 43, which noted that Japanese faces were "much fairer" than those of the Chinese, and travel-writer Bird's unhesitating and condescending references to the Japanese as a yellow people: *Unbeaten Tracks in Japan*, 12, 255. See also Kowner, "Lighter than Yellow, But Not Enough."

CHAPTER 2
Taxonomies of Yellow: Linnaeus, Blumenbach, and the Making of a "Mongolian" Race in the Eighteenth Century

1. Bodin, *Les six livres*, 526, 536–37 (*Six Bookes*, 554, 562). For a review of climatological theories before the end of the eighteenth century see Thomas, *The Environmental Basis of Society*, 30–75.

2. On the problem of Asia as a continent see Lewis and Wigen, *The Myth of Continents*, 47–103.

3. Bruno, *Opera latine*, ed. Fiorentino, 1:2:282. An English translation of the passage is given in Slotkin, *Readings in Early Anthropology*, 43. A recent Italian translation revealingly translates *fulva* as *rossa*, thus incorrectly aligning it with modern prejudices about "red" Native Americans (Bruno, *Opere latine*, ed. Monti, 783). On the designation "Yellow Carib" in the eighteenth century see Hulme, "Black, Yellow, and White on St. Vincent." According to Horn's *Arca Noae* of 1666, the sons of Shem, who went to India, were *flavos* (i.e., yellow); reminiscent of the Song of Songs, Horn also called them *fusci & pulchri* (dark and beautiful) (37–38).

4. Bernier, "Nouvelle division de la Terre." For an English translation see Bendyshe, "The History of Anthropology," 360–64. See also Bernier, *Histoire de la dernière révolution des états du Grand Mogol*; in English as *The History of the Late Revolution of the Empire of the Great Mogol*. On Bernier and his influence, see Stuurman, "François Bernier and the Invention of Racial Classification." On the development of the term "race" see Smedley, *Race in North America*, 36–40; and Hudson, "From 'Nation' to 'Race.'"

5. Bernier, "Nouvelle division de la Terre," 148; Leibniz, *Otium hanoveranum*, 37–38, 160; Leibniz, *Sämtliche Schriften und Briefe*, 13:544–45. On Leibniz, race, and language see Fenves, "Imagining an Inundation of Australians."

6. Bernier, "Nouvelle division de la Terre," 149.

7. Bernier, "Nouvelle division de la Terre," 152–53; Montesquieu, *Oeuvres complètes*, ed. Callois, 1:272 (letter 96); Voltaire, *Traité de métaphysique*, 5; Nicolle de la Croix, *Géographie moderne*, 1:62; Kant, in Engel, *Der Philosoph für die Welt*, 2:138, 153, 160; Kant, "Bestimmung des Begriffs einer Menschenrace," 393–94; Kant, *Physische Geographie*, in *Werke*, 9:316. See also

Dampier, *New Voyage* (1:326, 407, 2:1:40), where Indians were frequently called yellow as opposed to the "ashy" Chinese. There is also a *Pensée* of Montesquieu (*Oeuvres complètes*, ed. Masson, 2:29), not published during his lifetime, that referred to "les peuples jaunes d'Asie"; this also seemed to be an allusion to India and not China.

8. Bernier, "Nouvelle division de la Terre," 149–51.

9. Bernier, "Nouvelle division de la Terre," 151–53.

10. According to Dikötter, *The Discourse of Race in Modern China*, 55, Bernier's text was "the first scientific work in which the notion of a 'yellow race' appeared." This was then repeated in Dikötter, "Racial Discourse in China," 12. The source is Huard, "Depuis quand avons-nous la notion d'une race jaune?" 42; and repeated in Huard, *Chinese Medicine*, 115.

11. Linnaeus, *Systema naturae*, unnumbered folios; Lovejoy, *The Great Chain of Being*.

12. According to Bradley there were five "sorts" of human being including "the *White-Men*, which are *Europeans*, who have *Beards*," a "Sort of *White-Men* in *America* (as I am told) who only differ from us in having no *Beards*," "the *Mulattoes*, which have their *Skins* almost of a *Copper* Colour, small *Eyes*, and *strait black Hair*," "the *Blacks*, who have *strait black Hair*," and "the *Blacks* of *Guinea*, whose *Hair* is curled, like the *Wool* of a *Sheep*" (*A Philosophical Account of the Works of Nature*, 231). Linnaeus's source for "red" Americans was also unclear, future stereotypes notwithstanding: Shoemaker, *Strange Likeness*, 130.

13. André, *Termes de couleur*, 123–25.

14. Beazley, *Dawn of Modern Geography*, 2:549–642; Miller, *Mappaemundi*, esp. 1:53; Horn, *De originibus americanis*, 200; Sande, *De missione legatorum japonensium*, 407–8.

15. Mendoza, *Historia . . . del gran reyno de la China*, 5; Mendoza, *Rerum morumque in regno Chinensi*, 5 (although an earlier Latin edition [*Nova et succincta . . . historia de . . . regno China*, 32] had chosen *subflavescunt* [i.e., yellowish] instead!); Ricci, *De Christiana expeditione*, 86.

16. *Albus* appeared in the second edition (1740), *rufescens* in the sixth (1748), and *niger* in the third (1740), which also included the German *gelblich*.

17. Linnaeus, *Systema naturae*, 10th ed., 1:21.

18. André, *Termes de couleur*, 137–38; Ernout and Meillet, *Dictionnaire étymologique de la langue latine*, 1:660. A similar statement was made by Telesio in his *Libellus de coloribus* of 1528 (*Opera*, 182); see also Osborne, "Telesio's Dictionary of Latin Color Terms," 144.

19. For instance Turner, *De morbis cutaneis*, 95–102; Bickerton, *Accurate Disquisitions in Physick*, 95–99; and Ball, *The Modern Practice of Physic*, 2:205–12. See also *A Source Book in Medieval Science*, 707, for a ninth-century Galenic text on the effect of humors on skin color.

20. Mandeville, *Mandeville's Travels*, 2:231, see also 1:116; Kant, in Engel, *Der Philosoph für die Welt*, 2:153–54, and repeated in Mellin, *Encyclopädisches Wörterbuch der kritischen Philosophie*, 4:751. A similar statement was given in Blumenbach, *De generis humani varietate nativa*, 3rd ed., 131, where it was suggested that the whites of the eyes of *fuscus* nations—especially Indians, Americans, and Ethiopians—were tinged with a jaundiced yellow (*Anthropological Treatises*, 213). Kant credited Ives, *Voyage from England to India in the Year 1754*, for this information, but I have been unable to locate it in either the English or the German (*Reisen nach Indien und Persien*) editions of Ives's narrative.

21. Linnaeus, *Systema naturae*, unnumbered folio; Linnaeus, *Fundamenta botanica*, 35; Linnaeus, *Philosophia botanica*, 31, 279; Linnaeus, *Clavis medicinae*, 24. I am grateful to Karen Reeds for this last reference. See also Cain, "Linnaeus's Natural and Artificial Arrangements of Plants." Rogers links chlorosis to the medical condition familiarly known as "green sickness" (*Colors of Mankind*, 31).

22. Linnaeus, *Philosophia botanica*, 244; on Linnaeus and the seductive East see Koerner, *Linnaeus*, 95–112.

23. Pastoureau, *Figures et couleurs*; Pastoureau, *Couleurs, images, symboles*; Pastoureau, "Formes et couleurs du désordre"; Ellis, "The Psychology of Yellow"; *Handwörterbuch des deutschen Aberglaubens*, 3:570–83.

24. Goethe, *Werke*, 13:495–96. See also *Goethe's Color Theory*, 168–69.

25. Linnaeus, *Systema naturae*, 10th ed., sig. A–Av; Osbeck, *Dagbok öfwer en ostindisk resa*, 168 (in English, *A Voyage to China and the East Indies*, 1:266; in German, *Reise nach Ostindien und China*, 219); Torén's *Ostindisk resa til Suratte* appeared in the same volume (358 [English, 2:233; German, 489]). There is also a French translation, *Voyage de Mons. Olof Torée*, 67.

26. Torén, in Osbeck, *Dagbok*, 373 (English, 2:261; German, 510; *Voyage de Mons. Olof Torée*, 88).

27. Linnaeus, *Systema naturae*, 12th ed., 1:29; Linnaeus, *Vollständiges Natursystem*, 1:89–98. In 1860, Geoffroy Saint-Hilaire also translated *luridus* as "dark"—in this case, *basané* ("Sur la classification anthropologique," 126).

28. Linnaeus, *Systema naturae*, 13th ed., 1:23–24. A 1792 English translation of this edition added such terms as "sooty," "tawny," "yellowish brown," and "dark blackish brown" (Linnaeus, *The Animal Kingdom*, 45–46); another, fuller translation appeared in 1806, which also chose "sooty" to describe the color of Asians (Linnaeus, *A General System of Nature*, 1:9).

29. Blumenbach, *De generis humani varietate nativa*, 3rd ed., 297 (*Anthropological Treatises*, 267).

30. Buffon, "Variétés dans l'espèce humaine," in *Histoire naturelle, générale et particulière*, 3:371–530. "Nature in all her perfection," he remarked, "has made men white" (3:503). For an English translation see Buffon, *Natural History, General and Particular*, 3:57–207.

31. Buffon, *Histoire naturelle*, 3:379–93, 403–4 (*Natural History*, 3:65–81, 89–90). In addition to Pslamanazar, Buffon also cited such fictionalized accounts as the *Voyage* of Saunier de Beaumont, whose pseudonym was Inigo Biervillas.

32. Buffon, *Histoire naturelle*, 9:2 (*Natural History*, 5:64).

33. Blumenbach, *De generis humani varietate nativa*, 40–42 (*Anthropological Treatises*, 98–99).

34. Blumenbach, *Handbuch der Naturgeschichte*, 63; *De generis humani varietate nativa*, 2nd ed., 51–52 (*Anthropological Treatises*, 99–100).

35. Bernasconi, "Who Invented the Concept of Race?"

36. Kant, *Von den verschiedenen Racen der Menschen*, 4–6, 11. For an English translation see *Race and the Enlightenment*, 38–48.

37. Kant, in Engel, *Der Philosoph für die Welt*, 2:138, 159–60.

38. Kant, "Bestimmung des Begriffs einer Menschenrace," 393, 408; Pallas, *Sammlungen historischer Nachrichten über die mongolischen Völkerschaften*, 1:99. For modern commentary see Adickes, *Kant als Naturforscher*, 2:413–14; Bindman, *Ape to Apollo*, 163–89; and Zammito, "Policing Polygeneticism in Germany," 42.

39. Nicolle de la Croix, *Géographie moderne*, 1:62–63, divided the white race into white, brown, yellow, and olive varieties (Indians were yellow; Chinese and Japanese white). Pownall found three varieties (white, red, and black), corresponding to the sons of Noah. The "red" variety included the Tartars (*The Administration of the Colonies*, 155–57; Blumenbach cited this from *A New Collection of Voyages, Discoveries, and Travels*, 2:273–74). Goldsmith said there were six: deep brown Laplanders, black Africans, olive Tartars, olive Indians, red or copper Americans, and fair and beautiful Europeans (*History of the Earth*, 2:212–30). Erxleben also claimed six varieties, including Laplanders, Tartars who were olive, "Asiatics" who were *luridus* (!), Europeans, Africans, and Americans (*Systema regni animalis*, 1–2). Zimmermann decided upon four geographical areas, with the "original" people coming from Western Asia / Scythia; color was not emphasized: *Geographische Geschichte des Menschen*, 1:114–17.

40. Blumenbach, *De generis humani varietate nativa*, 3rd ed., 79–80, 118 (*Anthropological Treatises*, 193, 208). An obscure exception was Mitchell, "An Essay upon the Causes of the Different Colours of People in Different Climates," 146–47, who argued that originary man was "tawny," and that both Europeans and Africans had "degenerated" from that color.

41. Gould, *The Mismeasure of Man*, 401–12; Junker, "Blumenbach's Racial Geometry."

42. Blumenbach, *De generis humani varietate nativa*, 3rd ed., 119–22, 286 (*Anthropological Treatises*, 209–10, 264–65); at 299 he also linked *gilvus* to the French *jaunâtre*. The German translation used both *gelb* and *gelbbraun* (*Über die natürlichen Verschiedenheiten im Menschengeschlechte*, 95, 206). In

Blumenbach's *Beyträge zur Naturgeschichte* of 1790 Mongolians were *gelbbraun* (82), later changed to *waizengelb* (i.e., wheaten yellow) in the second edition of 1806 (71). Horn's *Arca Noae* described the Jews as "the color of boxwood" (37); see also Braude, "Sons of Noah," 112 n. 8; and Goldenberg, *The Curse of Ham*, 95. We have already discussed quinces in chapter 1.

43. Even in Blumenbach's *Beyträge zur Naturgeschichte* of 1790 they did not appear. The classic source for "Caucasian" (or, rather, "Georgian") beauty was a statement in Chardin's *Voyages*, 1:128, first published in 1686 and then definitively in 1711; see also Blumenbach, *De generis humani varietate nativa*, 3rd ed., 303 (*Anthropological Treatises*, 269); and Augstein, "From the Land of the Bible to the Caucasus and Beyond."

44. Blumenbach, *De generis humani varietate nativa*, 3rd ed., 304–7; Blumenbach, *Beyträge zur Naturgeschichte*, 2nd ed., 72 (*Anthropological Treatises*, 269–70, 304). Two contemporary travelers who remarked on the Tartars were Charpentier de Cossigny, *Voyage à Canton*, 280 (who called them *plus blancs* than the Chinese); and Turner, *An Account of an Embassy to the Court of the Teshoo Lama*, 247 (who said of one man that "his complexion was not darker than that of an Arab or a Spaniard").

45. Blumenbach, *De generis humani varietate nativa*, 65 (*Anthropological Treatises*, 119), and repeated in Blumenbach, *De generis humani varietate nativa*, 2nd ed., 84.

46. On Chinese visitors see Appleton, *Cycle of Cathay*, 130–36. See also Ong, "Wang-y-Tong."

47. Jones, "On the Second Classical Book of the Chinese"; Jones, *Works*, 1:365–73. Whang was also mentioned in a letter from Jones from 1785 (*Letters*, 2:684–85). See also Fan, "Sir William Jones's Chinese Studies."

48. Blumenbach, *Correspondence*, 1:18; Nichols, *Illustrations*, 5:318–19; Chambers, *Treatise of Oriental Gardening*, 2nd ed., 115–16 (and in French as *Discours . . . par Tan Chet-Qua de Quang-Cheou-Fou*, 7–8).

49. Lichtenberg, "Von den Kriegs- und Fast-Schulen der Schinesen," *Vermischte Schriften*, 5:247–48; reprinted in Hsia, *Deutsche Denker über China*, 107.

50. Blumenbach, *De generis humani varietate nativa*, 65; 2nd ed., 84; 3rd ed., 176–223 (*Anthropological Treatises*, 119, 227–43). On flat faces among the Chinese see Buffon, *Histoire naturelle*, 3:387, and on Africans, 3:458–59 (*Natural History*, 3:72, 142), a stereotype that also made an appearance in Swift's *Gulliver's Travels*. See also Meijer, *Race and Aesthetics in the Anthropology of Petrus Camper*, 154–58. Others who repeated the theory were Osbeck, *Dagbok*, 171; and Torén, in Osbeck, *Dagbok*, 358. The Chinese skulls were briefly described in Buffon, *Histoire naturelle*, 14:377.

51. Blumenbach, *De generis humani varietate nativa*, 3rd ed., 179–80, 194, 206–7 (*Anthropological Treatises*, 228, 233, 237).

52. Examples include Staunton, *An Authentic Account of an Embassy . . . to the Emperor of China*, 2:47; Van Braam Houckgeest, *Voyage . . . vers l'empereur de la Chine*, 1:72 (*An Authentic Account of the Embassy . . . to the Court of the Emperor of China*, 1:95); Guignes, *Voyages à Peking*, 2:159; and Abel, *Narrative of a Journey in the Interior of China*, 78.

CHAPTER 3

Nineteenth-Century Anthropology and the
Measurement of "Mongolian" Skin Color

1. Stepan, *The Idea of Race in Science*, 12–19; Jackson and Weidman, *Race, Racism, and Science*, 41–42; Augstein, "Land of the Bible," 64–72.
2. Harris, *The Rise of Anthropological Theory*, 83–94.
3. Oken, *Allgemeine Naturgeschichte*, 7:1852.
4. Polo, *The Book of Ser Marco Polo*; *The Texts and Versions of John de Plano Carpini and William de Rubruquis*.
5. *Texts and Versions of John de Plano Carpini*, 74–144. On early confusion between Mongol and Tartar see Saunders, "Matthew Paris and the Mongols," 124 n. 16; Bezzola, *Die Mongolen in abenländischer Sicht*, 125–26; and Klopprogge, *Ursprung und Ausprägung des abendländischen Mongolenbildes*, 155–59. Early eighteenth-century sources include Avril, *Voyage en divers états d'Europe et d'Asie* (in English as *Travels into Divers Parts of Europe and Asia*); *Histoire généalogique des Tatars* (in English as *A General History of the Turks, Moguls, and Tatars*); Witsen, *Noord en Oost Tartarye*; Pétits de la Croix, *Histoire du grand Genghizcan* (in English as *The History of Genghizcan the Great*); Strahlenberg, *Das nord- und ostliche Theil von Europa und Asia* (in English as *An Histori-Geographical Description of the North and Eastern Part of Europe and Asia*); *Relation de la Grande Tartarie* (also published in *Receuil de voyages au nord*, 10:1–301); Gaubil, *Histoire de Gentchiscan*; and Guignes, *Histoire générale des Huns*. See also Baddeley, *Russia, Mongolia, China*. In the 1770s Gibbon's *Decline and Fall of the Roman Empire* continued to refer to northern "pastoral nations" as Tartar (1:899–956).
6. Both "China" and "Japan" as standard terms date from the sixteenth century. In the case of Ortelius it was not until around 1600 that China and Japan had been given separate entries, and in earlier editions one can actually trace a gradual increase of information. A similar pattern can be found in editions of Heylyn's *Microcosmos*, first published in 1621 and later vastly expanded as *Cosmographie* in 1652. Convenient summaries of medieval and early modern knowledge about China and Japan appear in Lach and Van Kley, *Asia*, 1:1–86; Yule, *Cathay and the Way Thither*, 1:1–182; and Boxer, *Christian Century in Japan*, 1–40.

7. Georgi's text appeared in English as *Russia; or, a Compleat Account of All the Nations Which Compose That Empire*; see also his *Bemerkungen einer Reise im russischen Reich*; as well as Pallas, *Reise durch verschiedene Provinzen des russischen Reichs*. Blumenbach noted that the Mongols' "inhuman" reputation had been erroneously attributed to the Tartars since the thirteenth century (*De generis humani varietate nativa*, 3rd ed., 305–6 [*Anthropological Treatises*, 269–70]). See also Cecere, "Russia and its 'Orient.'" By 1802, Pinkerton could remark that "the vague name of Tartary is nearly discarded": *Modern Geography*, 2:44.

8. Bindman, *Ape to Apollo*, 219–21.

9. Meiners, *Grundriß der Geschichte der Menschheit*, 43, 89. Regarding the newly "westernized" nature of the Tartars as opposed to the Mongols, in 1729 the English translator of the *Histoire généalogique des Tatars* even noted that England was "no[ne] other than a Colony of Tatars" (*A General History of the Turks, Moguls, and Tatars*, 1:iii).

10. Meiners, *Grundriß*, 47; 2nd. ed., 93.

11. On their debate see Dougherty, "Christoph Meiners und Johann Friedrich Blumenbach."

12. Malthus, *An Essay on the Principle of Population*, 39, 49.

13. *Lettres édifiantes et curieuses*, 24:64. The statement was then repeated by Buffon, *Histoire naturelle*, 3:388–89 (*Natural History*, 3:75).

14. Boemus, *Omnium gentium mores*, 163; Münster, *Cosmographia*, 1059; Tavernier, *Six voyages*, 1:330, 2:501 (*Six Voyages* [English ed.], part 1, p. 127, part 2, p. 204); Buffon, *Histoire naturelle*, 3:381 (*Natural History*, 3:68). Buffon elsewhere noted that Africans, just like Europeans, had both their Tartars and their Circassians—that is, their ugly and beautiful peoples (*Histoire naturelle*, 3:453 [*Natural History*, 3:137]).

15. Meijer, *Race and Aesthetics*, 139–44.

16. Herder, *Sämtliche Werke*, 13:215–17, 14:13 (*Outlines of a Philosophy of the History of Man*, 137–39, 296). See also Lach, "China in Western Thought and Culture," 1:366; Blue, "China and Western Social Thought in the Modern Period"; Jones, *The Image of China in Western Social and Political Thought*, 67–98; and Dawson, *The Chinese Chameleon*, 65–89. Similarly, in 1795 Condorcet would describe the first epoch of human progress as an organization of people into tribes or *peuplades*, rendered into English as "hordes." The Chinese, he added, "seem to have preceded all others in the arts and sciences, only to see themselves successively eclipsed by them all": *Esquisse d'un tableau historique des progrès de l'esprit humain*, 21, 70 (*Outlines of an Historical View of the Progress of the Human Mind*, 21, 66).

17. Blumenbach, *De generis humani varietate nativa*, 3rd ed., 320 (*Anthropological Treatises*, 275).

18. Brownish red and yellowish: Duméril, *Zoologie analytique*, 7; yellow and olive: Cuvier, *Le règne animal*, 1:94–95 (*The Animal Kingdom*, 1:97);

brownish yellow: Virey, *Histoire naturelle*, 1:459; yellow orange and pistachio green: Desmoulins, *Histoire naturelle des races humaines*, 164. For yellow Indians see Mellin, *Encyclopädisches Wörterbuch*, 4:745; and Girtanner, *Über das kantische Prinzip für die Naturgeschichte*, 59. Smith also referred to northern Indians as yellow: *An Essay on the Causes of the Variety of Complexion*, 64.

19. Virey, *Histoire naturelle*, 1:437–38; Malte-Brun, *Précis de la géographie universelle*, 2:554 (*Universal Geography*, 1:255); Rudolphi, *Grundriss der Physiologie*, 1:57, 62.

20. Bory de Saint-Vincent, *L'homme*, 1:75–83; Duméril, *Zoologie analytique*, 6–7; Cuvier, *Le règne animal*, 1:94–100 (*Animal Kingdom*, 1:96–103); Virey, *Histoire naturelle*, 1:437–38; review of Desmoulins in *Bulletin des sciences naturelles et de géologie*.

21. Rémusat, *Recherches sur les langues tartares*, xxxvi; Rémusat, *Nouveaux mélanges asiatiques*, 1:31. Thirty-seven varieties were suggested by Girtanner, *Über das kantische Prinzip*, 59–73.

22. Mellin, *Encyclopädisches Wörterbuch*, 4:743. See also McLaughlin, "Kant on Heredity and Adaptation."

23. Cuvier, *Le règne animal*, 1:95 (*Animal Kingdom*, 1:97); Lawrence, *Lectures on Physiology, Zoology, and the Natural History of Man*, 483.

24. Lawrence, *Lectures on Physiology*, 478, 483–84.

25. Lawrence, *Lectures on Physiology*, 483, 530. An exception was Van Amringe, who rejected the category "Mongolian"; the Chinese, he said, did not derive from such a "nation of pastoral wanderers and robbers," although their physical features were the same: *An Investigation of the Theories of the Natural History of Man*, 69–70.

26. Prichard, *Researches*, 545; 3rd ed., 1:1–2; *Natural History of Man*, 230. The notion that the only attractive Chinese were descendants of foreigners appeared in the first European book on China (Cruz, *Tractado*, 73 [Boxer, *South China*, 137]). It was repeated in, among other texts, Lucena, *Francisco de Xavier*, 861.

27. Carus, *Denkschrift zum hundertjährigen Geburtsfeste Goethe's*, 14–15; Klemm, *Allgemeine Cultur-Geschichte der Menschheit*, 6:5; Smith, *The Natural History of the Human Species*, 285; Bachman, *The Doctrine of the Unity of the Human Race Examined*, 31; Latham, *The Natural History of the Varieties of Man*, 2, 14.

28. Bachman, *Doctrine of Unity*, 291–92.

29. Morton, *Crania Americana*; Morton, "Observations on the Size of the Brain in Various Races." See also Gould, *Mismeasure of Man*, 50–69.

30. Nott and Gliddon, *Types of Mankind*, 448–50.

31. Gobineau, *Essai sur l'inégalité des races humaines*, 1:247, 352, 363–64; 2:247–48, 349 (*The Moral and Intellectual Diversity of Races*, 450, 459–60; volume 2 not translated). A strange twentieth-century variant appeared in Legendre, *La civilisation chinoise moderne*, who similarly argued that

Chinese civilization was a white import; the "yellow race," he claimed, did not even exist (218–38).

32. The Société ethnologique de Paris had been founded in 1839, followed by the Ethnological Society of New York in 1842, the Ethnological Society of London in 1843, and the Koninklijk Instituut voor Taal-, Land- en Volkenkunde in Holland in 1851.

33. Schiller, *Paul Broca*, 163.

34. This went through six editions and, beginning with the second edition of 1892, was published by the Royal Anthropological Institute in London as *Notes and Queries on Anthropology*. It was reprinted as late as 1971. Competing guides published in France and Germany include "Instruction générale adressée aux voyageurs"; *Instructions générales aux voyageurs*; and *Anleitung zu wissenschaftlichen Beobachtungen auf Reisen*.

35. Broca, *Instructions générales pour les recherches et observations anthropologiques*, 134–36. The text was also reprinted in the 1865 *Archives de médecine navale*.

36. Broca, *Instructions générales*, 54.

37. An exception was Yule, "On the Influence of Bias and of Personal Equation in Statistics of Ill-Defined Qualities."

38. Topinard, *L'anthropologie*, 364 (*Anthropology*, 344); Broca, *Recherches sur l'hybridité animale*, 463; Broca, *Instructions générales*, 55.

39. Broca, "Instructions générales pour les recherches anthropologiques"; Broca, *Instructions générales*, 54–55.

40. Broca, "Échelle chromatique des yeux," 592–93; Broca, "Tableau chromatique des yeux, de la peau et des cheveux," 768–73; Broca, *Instructions générales*, 45.

41. Tylor, *Anthropology*, 69–71; Beddoe, *The Races of Britain*, 1–8; Beddoe, "Observations on the Natural Colour of the Skin in Certain Oriental Races," 258, 261.

42. Topinard, *Éléments*, 317; *Notes and Queries on Anthropology*, 64; Hrdlicka, *The Old Americans*, 22; *Hrdlicka's Practical Anthropometry*, 107. At first, Hrdlicka's *Directions for Collecting Information and Specimens for Physical Anthropology* retained thirty-two of Broca's original thirty-four colors, although they had also been rearranged "to facilitate determination" (25).

43. Hrdlicka, "Anthropometry," 46–67, 290, 314; *Hrdlicka's Practical Anthropometry*, 107; Deniker, *Les races et les peuples de la terre*, 64 (*The Races of Man*, 47). See also Martin, *Lehrbuch der Anthropologie*, 206; and Comas, *Manual of Physical Anthropology*, 264–65.

44. Luschan, "Einige wesentliche Fortschritte in der Technik der physischen Anthropologie"; Luschan, "Über Hautfarbentafeln"; Virchow, "Von Luschan'schen Farbentafel"; Fritsch, "Die Feststellung der menschlichen Hautfarben"; Fritsch, "Bemerkungen zu der Hautfarbentafel"; Martin,

Lehrbuch der Anthropologie, 206–7; Hrdlicka, "Anthropometry," 290; Thomson, "On the Treatment and Utilization of Anthropological Data," 27; Ranke, "Über die Hautfarbe der südamerikanischen Indianer"; Sarasin and Sarasin, *Ergebnisse naturwissenschaftlicher Forschungen auf Ceylon*, 3:91–96; Sarasin and Sarasin, *Materialien zur Naturgeschichte der Insel Celebes*, vol. 5, part 2, table 2; Gates, *Pedigrees of Negro Families*, 253–54.

45. Hintze, "Der Hautfarbenfächer und das Hautfarbendiagramm"; Fritsch, "Feststellung der menschlichen Hautfarben," 88; Hrdlicka, "Anthropometry," 313; Herskovits, *The Anthropometry of the American Negro*, 39; Basler, "Über eine Hautfarbentafel für Europäer."

46. Maxwell, *Scientific Papers*, 1:126–54; Bradley, *Color in the Schoolroom*, 52–59; Bradley, *Elementary Color*, 18–22; *Bradley's Kindergarten Material and School Aids*, 59. On Maxwell's experiments see Sherman, *Colour Vision in the Nineteenth Century*, 153–83; and Harman, *The Natural Philosophy of James Clerk Maxwell*, 37–48.

47. Strong, "A Quantitative Study of Variation in the Smaller North-American Shrikes."

48. Davenport, *Inheritance in Poultry*; Davenport, "Heredity of Skin Pigment in Man," 643, 663, 705. On Davenport's career see Kevles, *In the Name of Eugenics*, 41–56; and Barkan, *The Retreat of Scientific Racism*, 67–76.

49. Davenport, "Heredity of Skin Pigment," 643–44, 663; *The Heredity of Skin Color in Negro-White Crosses*, 1; Pearson, "Note on the Skin-Colour of the Crosses Between Negro and White," 349.

50. Davenport, *Negro-White Crosses*, 2; Sollors, *Neither Black Nor White Yet Both*, 137; Bowman, "The Color-Top Method of Estimating Skin Pigmentation," 59; Herskovits, *Anthropometry*, 39.

51. Davenport, "Heredity of Skin Pigment," 664.

52. Blakeslee and Warner, "Correlation Between Egg-Laying Activity and Yellow Pigment in the Domestic Fowl"; Harris, Blakeslee, and Warner, "The Correlation Between Body Pigmentation and Egg Production in the Domestic Fowl"; Palmer and Kempster, "The Influence of Specific Feeds and Certain Pigments on the Color of the Egg Yolk"; Wright, "The Effects in Combination of the Major Color-Factors of the Guinea Pig." Colored disks were also used for soil analysis: Bodman, "Color Discs Used in Soil Color Analysis."

53. Rowntree and Brown, "A Tintometer for the Analysis of the Color of the Skin"; Edwards and Duntley, "The Pigments and Color of Living Human Skin"; Byard, "Quantitative Genetics of Human Skin Color."

54. Davenport, "The Skin Colors of the Races of Mankind," 44, 46; Topinard, *L'anthropologie*, 363–64 (*Anthropology*, 344).

55. Luschan, "Über Hautfarbentafeln," 402. Bradley had originally employed N to stand for black because B was used for the blue disk instead; this

notation was retained by Davenport and others even though they no longer utilized the blue. See Pillsbury, "A New Color Scheme"; Bradley, "The Color Question Again"; and Bradley, *Elementary Color*, 19–20.

56. Davenport, *Heredity in Relation to Eugenics*, 37. Such "crosses" were not considered white even if "they might have mental and moral qualities as good and typically Caucasian" as their white ancestors (38).

57. Herskovits, *The American Negro*; Herskovits, *Anthropometry*, 18; Herskovits, "Does the Negro Know His Father?" 308; Barnes, "The Inheritance of Pigmentation in the American Negro"; Harris, "The San Blas Indians"; Davenport, "Notes on Physical Anthropology of Australian Aborigines"; Davenport, "Skin Colors of Races," 47.

58. Bradley, *Elementary Color*, 35; Blackwood, "Racial Differences in Skin-Colour as Recorded by the Colour Top," 163; Herskovits, *Anthropometry*, 73–74; Eickstedt, *Rassenkunde und Rassengeschichte der Menschheit*, 49–50; Bowman, "Color-Top Method," 60–61; Todd, Blackwood, and Beecher, "Skin Pigmentation," 189; Sumner, "Linear and Colorimetric Measurements of Small Mammals," 203; Todd and Van Gorder, "The Quantitative Determination of Black Pigmentation in the Skin of the American Negro," 247.

59. Todd and Van Gorder, "Quantitative Determination," 243–44; Harris, "San Blas Indians," 47.

60. Davenport, "Skin Colors of Races," 44–46. Herskovits agreed that the oxblood disk represented a "real" component of skin color (*Anthropometry*, 220).

61. Davenport, "Skin Colors of Races," 47–48; cf. Davenport, *Negro-White Crosses*, 28.

62. Blackwood, "Racial Differences," 144, 152, 160–65.

63. Sheard and Brown, "The Spectrophotometric Analysis of the Color of the Skin"; Shaxby and Bonnell, "On Skin Colour"; Hardy, "A New Recording Spectrophotometer"; Sheard and Brunsting, "The Color of the Skin as Analyzed by Spectrophotometric Methods"; Weiner, "A Spectrophotometer for Measurement of Skin Colour," 152.

64. Robins, *Biological Perspectives on Human Pigmentation*, 98–103; Dorno, "Beiträge zur Kenntnis des Sonnen- und Quarzlicht-Erythems und -Pigmentes," 74–75; Edwards and Duntley, "Pigments and Color," 6. In 1952 Gates noted that "the hemoglobin curve of [four Chinese subjects] is clear and the yellowish skin color may be due entirely to dilute melanin" ("Studies of Interracial Crossing," 34). In 1965 Coon remarked that Bushmen and "Mongoloids" had skins that "appear yellowish or yellowish brown" (*The Living Races of Man*, 234), and a few years later a small controversy arose over the true cause of "yellow oriental skin" (Daniels, Post, and Johnson, "Theories of the Role of Pigment in the Evolution of Human Races," 16). As recently as 1995 an essay on the origins of the "northern Mongoloids" took

their "pale yellow skin" for granted: Guthrie, "The Mammoth Steppe and the Origin of Mongoloids and Their Dispersal," 179.

65. Davenport, *Heredity in Relation to Eugenics*, 241–49. Davenport was infamous for his notion that even "love of the sea" was genetic; see Kevles, *Name of Eugenics*, 48–49.

CHAPTER 4
East Asian Bodies in Nineteenth-Century Medicine: The Mongolian Eye, the Mongolian Spot, and "Mongolism"

1. *Texts and Versions of John de Plano Carpini*, 76, 109; *Sinica franciscana*, 1:236; Ramusio, *Delle navigationi et viaggi*, 1:354v, 1:197v (misnumbered as 198v); Münster, *Cosmographia*, 1059; Cruz, *Tractado*, 73 (Boxer, *South China*, 137); Mendoza, *Historia . . . del gran reyno de la China*, 21; Couto, *Da Asia*, 13:265–66; Xavier, *Epistolae*, 2:291.

2. Ricci, *Fonti Ricciane*, 1:88; Ricci, *De Christiana expeditione*, 86; Martini, *Novus atlas sinensis*, 7; Nieuhof, *Het gezantschap*, part 2, p. 56.

3. Wood, *Life and Times*, 3:236; *Relatione . . . de' principi giapponesi*, sig. B3; Gualtieri, *Relationi della venuta*, 157; Lucena, *Francisco de Xavier*, 861; Du Jarric, *L'histoire des . . . Indes Orientales*, 733; Bartoli, *La Cina*, 67 n. 61; Cruz, *Tractado*, 73 (Boxer, *South China*, 137); Fróis, *Kulturgegensätze Europa-Japan*, 98 (this edition also includes the Portuguese original).

4. Bernier, "Nouvelle division de la Terre," 151 (Bendyshe, "History of Anthropology," 362); Kaempfer, *Geschichte*, 1:110 (*History*, 1:95); Thunberg, *Resa*, 3:279–80 (*Travels*, 3:251–52).

5. Blumenbach, *De generis humani varietate nativa*, 41; 3rd ed., 179 (*Anthropological Treatises*, 99, 265); cf. Blumenbach, *Über die natürlichen Verschiedenheiten im Menschengeschlechte*, 131. The second edition of *De generis humani varietate nativa* (51) referred to "narrow eyelids" only (*Anthropological Treatises*, 100).

6. Parry, *Journal of a Second Voyage for the Discovery of a North-West Passage*, 548; Finlayson, *Mission*, 229, 377; Siebold, *Nippon*, 1:299–303. Barrow, who traveled with the Macartney embassy to China in the 1790s, said of Chinese eyes that they were "depressed towards the nose" and "rounded in the corner next [to] the nose like the end of an ellipsis" (*Travels in China*, 49, 184).

7. Schön, *Handbuch der pathologischen Anatomie des menschlichen Auges*; Ammon, "Der Epicanthus"; Schön, "Zur Geschichte des Epicanthus." Ammon noted that the epicanthus made Europeans look more or less like Kalmucks (535). See also Lawrence, *A Treatise on the Diseases of the Eye*, 359.

8. Buffon, *Histoire naturelle*, 3:387 (*Natural History*, 3:72); Blumenbach, *De generis humani varietate nativa*, 65; 2nd ed., 85 (*Anthropological Treatises*, 119).

9. Siebold, *Nippon*, 2:306.

10. Sichel, "Mémoire sur l'épicanthus," 34. For other nineteenth-century sources see Onishi, "Honpojin no kenretsu (ganken haretsu)."

11. Chambers, *Vestiges of the Natural History of Creation*, 306–7; Serres, "Principes d'embryogénie," 763–65. Cf. Farrar, "Aptitudes of Races," 123, who noted that Chinese people represented "the best possible example of . . . arrested development" and "immobility."

12. Siebold, *Nippon*, 2:198; Ammon, "Der Epicanthus und das Epiblepharon," 345–46; Rossiianov, "Taming the Primitive," 217; Metchnikov [as Elias Metschnikoff], "Ueber die Beschaffenheit der Augenlider bei den Mongolen und Kaukasiern," 158–60. See also Deniker, "Étude sur les Kalmouks," 696–99; Ranke, "Ueber das Mongolenauge als provisorische Bildung bei deutschen Kindern"; and an essay by Richard Drews with the same title.

13. Chen, *Asian Blepharoplasty and the Eyelid Crease*, 273, where it is also claimed that all fetuses have such folds at first. See also Keith, *Human Embryology and Morphology*, 196; Bolk, "On the Origin of Human Races," 322–24; and Bolk, "Over mongolenplooi en mongoloide idiotie."

14. Siebold, *Nippon*, 1:301–2. To add to the confusion Drews ("Ueber das Mongolenauge als provisorische Bildung bei deutschen Kindern," 226–28) attempted to distinguish between a Mongolian eye and a Mongolian fold. See also Liu and Hsu, "Oriental Eyelids"; and Furukawa, "Aesthetic Surgery of Oriental Eyelids."

15. Pöch, "Zweiter Bericht . . . ," 119–22; Aichel, "Epicanthus, Mongolenfalte, Negerfalte, Hottentottenfalte, Indianerfalte." Sullivan, *Essentials of Anthropometry*, complained that the term was exceedingly confusing, so his guide provided illustrations of eyes that were "slanted," "slitted," and so on, but without the fold (51–53). Comas, *Manual*, 271–72, also tried to distinguish between three types of fold. For a useful review of this material see Chouke, "The Epicanthus or Mongolian Fold in Caucasian Children," 2–12. According to one researcher in 1897, however, even among East Asians the fold could disappear or diminish as the person aged: Iwanowski, "Zur Anthropologie der Mongolen," 69.

16. Baelz, *Erwin Bälz, das Leben eines deutschen Arztes im erwachenden Japan*; Baelz, *Awakening Japan*; Germann, *Ein Leben in Ostasien*. On Baelz's ethnic theories see Kowner, "Lighter than Yellow," 120–25.

17. Baelz, "Die körperlichen Eigenschaften der Japaner," 40, 93–95.

18. Baelz, "Referat über den Vortrag des Herrn . . . Dr. E. Baelz," 227, 234.

19. Baelz, "Menschen-Rassen Ost-Asiens mit specieller Rücksicht auf Japan," 188–89.

20. Baelz, "Zur Frage von der Rassen-Verwandtschaft zwischen Mongolen und Indianern"; Matignon, "Stigmates congénitaux et transitoires chez les Chinois"; Chemin, "Taches congénitales de la région sacro-lombaire";

Chemin, "Note sur les taches congénitales de la région sacro-lombaire chez les Annamites."

21. Brennemann, "The Sacral or So-Called 'Mongolian' Pigment Spots of Earliest Infancy and Childhood"; Deniker, "Les taches congénitales dans la région sacro-lombaire," 277.

22. Baelz, "Menschen-Rassen," 188; Baelz, "Noch einmal die blauen 'Mongolen-Flecke,'" 329.

23. Adachi, "Hautpigment beim Menschen und bei den Affen," 102–12, 119–23; Adachi and Fujisawa, "Mongolen-Kinderfleck bei Europäern"; Fujisawa, "Sogenannter Mongolen-Geburtsfleck der Kreuzhaut bei europäischen Kindern"; El Bahrawy, "Über den Mongolenfleck bei Europäern." Japanese texts suggested that the spot had been caused by coitus during pregnancy, by the *obi* worn by the pregnant mother over her kimono, by contact between the fetus and the placenta, and so on. Other cultures in which the spots appeared had analogous explanations; see ten Kate, "Die blauen Geburtsflecke"; ten Kate, "Neue Mitteilungen über die blauen Geburtsflecken"; and Brennemann, "Mongolian Pigment Spots," 25–26.

24. Baelz, "Mongolen-Flecke," 331.

25. Deniker, "Les taches congénitales"; Bloch, "Preuves ataviques de la transformation des races"; Ashmead, "The Mulberry-Colored Spots on the Skin of the Lower Spine of Japanese and Other Dark Races," 212; Epstein, "Über den blauen Kreuzfleck und andere mongoloide Erscheinungen bei europäischen Kindern," 62; Le Dantec, *Précis de pathologie exotique*, 2:954.

26. Hansen, "Bidrag til Vestgrønlaendernes Anthropologi," 237; Hansen, "Bidrag til Østgrønlaendernes Anthropologi," 38–39; Bloch, "Preuves ataviques," 618–19; Wardle, "Evanescent Congenital Pigmentation in the Sacro-Lumbar Region," 418.

27. Gumilla, *El Orinoco illustrado*, 1:82. A similar suggestion was made by Saabye, *Greenland*, 282; and Eschricht, *Zoologisch-anatomisch-physiologische Untersuchungen über die nordischen Wallthiere*, 70.

28. Ashmead, "Mulberry-Colored Spots," 213–14; Ashmead, "Relation of Syphilis with Japanese Racial Peculiarities and Customs," 395; cf. Toldt, "Über Hautzeichnung bei dichtbehaarten Säugetieren."

29. Mayerhofer, "Gegen die Mongolentheorie der sog. 'Mongolenflecke'"; Baur, Fischer, and Lenz, *Menschliche Erblichkeitslehre*, 98; Baur, Fisher, and Lenz, *Menschliche Erblehre*, 119–20.

30. Rivet, *Les origines de l'homme américain*, 68–71; Ratsimamanga, "Tache pigmentaire héréditaire et origines des Malgaches." One can still find claims about their racial import, as in a 2007 essay on the skin color of the ancient Egyptians that suggested that the spot was a genetic trace of the invasions of Attila the Hun (Reitz, "Die Hautfarbe der alten Ägypter," 164).

31. Larsen and Godfrey, "Sacral Pigment Spots," 256; Smialek, "Significance of Mongolian Spots"; Bittner and Newberger, "Pediatric Understanding

of Child Abuse and Neglect"; Dungy, "Mongolian Spots, Day Care Centers, and Child Abuse"; Asnes, "Buttock Bruises = Mongolian Spot."

32. Down, "Observations on an Ethnic Classification of Idiots"; and reprinted in *The Origins of Modern Psychiatry*, 15–18.

33. Shuttleworth, "The Physical Features of Idiocy."

34. Gould, *The Panda's Thumb*, 160–68; Down, "Observations," 260–61.

35. Huxley, "On the Distribution of the Races of Mankind"; Huxley, "On the Geographical Distribution of the Chief Modifications of Mankind"; Fraser and Mitchell, "Kalmuc [sic] Idiocy"; "Case and Autopsy of a Kalmuc [sic] Idiot," 162.

36. Down, "Observations," 260–61; Chambers, *Vestiges*, 289, 302, 307; Zihni, "The History of the Relationship Between the Concept and Treatment of People with Down's Syndrome," 74–76; Zihni, "Imitativeness and Down's Syndrome."

37. Booth, "Labels and Their Consequences"; Jackson, "Changing Depictions of Disease"; Rafter, *The Criminal Brain*.

38. Langdon-Down, "Some Observations on the Mongolian Type of Imbecility," 188–89; Shuttleworth, "Mongolian Imbecility," 662; Brousseau, *Mongolism*; Benda, *Mongolism and Cretinism*; Lowe, "The Eyes in Mongolism."

39. Tumpeer, "Mongolian Idiocy in a Chinese Boy"; Bleyer, "The Occurrence of Mongolism in Ethiopians"; Wagner, "Mongolism in Orientals."

40. Crookshank, *The Mongol in Our Midst*; Woodcock, "The History of the Medical Society of Individual Psychology"; Thomson, *Psychological Subjects*, 86.

41. Vogt, *Vorlesungen über den Menschen*, 2:284 (*Lectures on Man*, 467); Darwin, *The Descent of Man*, 1:228–31; Klaatsch and Hauser, "Homo Aurignacensis Hauseri"; Klaatsch, "Die Aurignac-Rasse und ihre Stellung im Stammbaum der Menschheit," 563–68; Klaatsch, "Menschenrassen und Menschenaffen," 92; Klaatsch, *Der Werdegang der Menschheit und die Entstehung der Kultur*, 90 (*The Evolution and Progress of Mankind*, 106).

42. Keith, "A New Theory of the Descent of Man" ("flimsy in the extreme"); Wells, *The Outline of History*, 1:63 ("These are very fanciful ideas, to be mentioned only to be dismissed"; in the revised edition of 1925, "very fanciful" was changed to "preposterous" [1:38]).

43. Kurz, "Zwei Chinesengehirne," 327; Kurz, "Das Chinesengehirn," 382. See also Kurz, "Der Unterkiefer des Chinesen"; Kurz, "Ergebnisse rassenanatomischer Untersuchungen über das Extremitätenskelet des Chinesen"; and Kurz, "Untersuchungen zur Anatomie der Weichteile beim Chinesen." Another follower of Klaatsch, also a great influence on Crookshank, was Sera, who divided human beings into six ape-types: "I caratteri della faccia e il polifiletismo dei primati." For extensive comparisons between orangutans and the Japanese, see also Sera, "Sul significato polifiletico delle differenze strutturali nell'arto inferiore di *Anthropoidea*."

44. Crookshank, *Mongol in Our Midst*, 2nd ed., 80; Crookshank, *Der Mongole in unserer Mitte*.

45. On the epicanthus: Crookshank, *Mongol in Our Midst*, 59, and strengthened in the 3rd ed., 143–47, where the author claimed to have seen such a fold on an orangutan. See also Sera, "I caratteri della faccia e il polifiletismo dei primati," 114. On the Mongolian spot: *Mongol in Our Midst*, 55–56; 3rd ed., 135–40. The quote is from the 1st ed., 9.

46. Schlapp, "Mongolism," 161; Herrman, "Mongolian Imbecility as an Anthropologic Problem," 528; Penrose, "The Blood Grouping of Mongolian Imbeciles"; Sapir, "The Race Problem," 40. See also Tredgold, *A Text-Book of Mental Deficiency*, 198–99, who wondered whether "this theory was ever seriously entertained by anyone but its author."

47. Crookshank, *Mongol in Our Midst*, 3rd ed., 38.

48. Crookshank, *Mongol in Our Midst*, 3rd ed., 134, 162, 165, 190. On the 1920s Mongol vogue see Childs, *Modernism and Eugenics*, 48–51, 90–91; and Bradshaw, "Eugenics," 52.

49. Allen et al., "Mongolism," 775; two letters of objection are printed as "Down's Syndrome," 935; Howard-Jones, "On the Diagnostic Term 'Down's Disease,'" 104; *Mongolism*, 88–90; Gibson, *Down's Syndrome*, xi. Cf. Merton's self-published *Mankind in the Unmaking*, which not only retained the term but also continued to insist on atavism and connections between Down syndrome patients and orangutans.

50. An example was Quatrefages, *Rapport sur les progrès de l'anthropologie*, 293–94, who mentioned "oblique Chinese eyes" that could be found among white French people, particularly women. This was also mentioned in the first edition of Crookshank, *Mongol in Our Midst*, 7, as an example of "mongoloid characteristics" in many Europeans.

51. Recent studies of Western medicine and East Asian bodies include Heinrich, *The Afterlife of Images*; Shah, *Contagious Divides*; and Rogaski, *Hygienic Modernity*.

52. Chambers, *Vestiges*, 305–8.

53. Chambers, *Vestiges*, 308.

54. Down, "Observations," 261; Baelz, "Die körperlichen Eigenschaften der Japaner," 39 (and on all "Mongolians": Baelz, *Die Ostasiaten*, 10–11); Hatch, "Some Studies upon the Chinese Brain," 101.

55. Gould, *Panda's Thumb*, 161.

CHAPTER 5
Yellow Peril: The Threat of a "Mongolian" Far East, 1895–1920

1. The Mongol khanate known as the Golden Horde, based perhaps on the color of their leaders' tents, could be considered one more element in this equation, although so far as I know it is never mentioned in connection

with the color of East Asian (or "Mongolian") skin. The term appears early: in Carpini's narrative as reported by Vincent of Beauvais in the middle of the thirteenth century (*Texts and Versions of John de Plano Carpini*, 100). See also Serruys, "Mongol Altan 'Gold' = 'Imperial.'" On the category "Yellow Mongol" see Baddeley, *Russia*, 2:60.

2. Gibbon, *Decline and Fall*, 1:899–956, 2:243–99; Lawrence, *Lectures on Physiology*, 483; Williams, *Middle Kingdom*, 1:39; Ratzel, *Die chinesische Auswanderung*, vi, 234–35; Quatrefages, *L'espèce humaine*, 411; Gobineau, *Oeuvres*, 3:xlvii; Blue, "Gobineau on China."

3. Thompson, *The Yellow Peril*. See also Gollwitzer, *Die gelbe Gefahr*; Laffey, *Imperialism and Ideology*, 70–94; Decornoy, *Péril jaune, peur blanche*; Mehnert, *Deutschland, Amerika und die "gelbe Gefahr"*; Miller, *The Unwelcome Immigrant*; McClellan, *The Heathen Chinee*; Markus, *Fear and Hatred*; Wu, *The Yellow Peril*; Blue, "Gobineau on China"; Lye, *America's Asia*; and Lyman, "The 'Yellow Peril' Mystique."

4. In the English-speaking world the image was given wide circulation in *Harper's Weekly* ("A Picture by Emperor William"); in the *Review of Reviews* for January 1896; and in Diosy, *The New Far East*, which appeared in several editions.

5. The German explication is cited in "The Far Eastern Situation from a German Standpoint"; a later notice in *The Times* also noted that the image did not seem to depict a dragon ("German Sympathies"). For the letters see Wilhelm II, *Briefe*, 291–95.

6. Wilhelm II, *Briefe*, 394; Cecil, *Wilhelm II*, 2:38; Röhl, *Wilhelm II*, 840–41.

7. See Pusey, *China and Charles Darwin*.

8. Tsu, *Failure, Nationalism, and Literature*, 43–44; Pusey, *China*, 147–8.

9. Some reformers were puzzled by the fact that Western science had classified the Chinese as part of a "Mongolian race"; on this subject and the Yellow Emperor generally see Dikötter, *Discourse of Race*, 67–71, 86–87; Chow, "Imagining Boundaries of Blood," 46–49; and Chow, "Narrating Nation, Race, and National Culture," 58–61.

10. Sun, *San min zhu yi*, in *Guo fu quan ji*, 1:5 (*The Three Principles of the People*, 8–9); Liu, "A Tentative Classification of the Races of China," 131.

11. Tsu, *Failure*, 43; on the Japanese cf. Clement, *A Handbook of Modern Japan*, 45: "It is well known that the Japanese are classed under the Mongolian (or Yellow) Race. They themselves boastfully assert that they belong to the 'golden race,' and are superior to Caucasians, who belong to the 'silver race.'"

12. Tsu, *Failure*, 79, 88–97; Pusey, *China*, 68, 98–99, 118, 315; Lehmann, *The Image of Japan*, 149–50; Gulick, *The White Peril in the Far East*; France, *Sur la pierre blanche*.

13. A modern lifestyle and culture journal begun in 1993 and published in Zhengzhou, capital of Henan province, is titled *Huanghe huangtu huangzhong ren* (Yellow River, Yellow Earth, Yellow Race).

14. Ricci, *China*, 66, 305; Martini, *Novus atlas sinensis*, 12, 14; Kao, "Essai sur l'antiquité des Chinois," 1:104–5; Gaubil, *Traité de la chronologie Chinoise*, 17:6–7.

15. Brosses, "Essai de géographie étymologique sur les noms donnés aux peuples Scythes anciens & modernes," 498–99. On *seres* see Pauly, *Paulys Realencyclopädie der classischen Altertumswissenschaft*, Zweite Reihe, vol. 2, cols. 1678–83; Yule, *Cathay*, 1:1–34; and Myers, "Marvels, Myths, and Misconceptions."

16. Amiot, "L'antiquité des Chinois, prouvée par les monuments," 2:175.

17. Topinard, *Éléments*, 63. On the Five Elements see Needham, *Science and Civilisation*, 2:232–73. It was not so long ago that boxes of crayons included a color called "flesh." See also Kornerup and Wanscher, *Methuen Handbook of Colour*, which defines "flesh" as "an average flesh color," "the color of the flesh of the Caucasian race" (161).

18. Klaproth, *Mémoires relatifs à l'Asie*, 3:472. In an essay on colors first published in 1784, Saint-Pierre noted that in China yellow was prized and that in India and China, among other places, white was color of the devil because it "contrasts sharply with the black color of these peoples. The Indians are black; the southern Chinese have very tanned (*basanée*) skin": *Études de la nature*, 2:84, 90–92 (*Studies of Nature*, 2:291, 298 ["much sun-burnt"]).

19. Zhao, *Zhufan zhi*, 20, 29, 31, 40, 56 (*Chau Ju-Kua: His Work . . . Entitled Chu-fan-chi*, 72, 84, 88, 103, 134).

20. See also Jones, *Yellow Music*.

21. Couto, *Da Asia*, 13:263. See also Ferguson, "Letters from Portuguese Captives," 437, 442; Boxer, *Christian Century in Japan*, 24–25; and Abramson, "Deep Eyes and High Noses." In 1602, one missionary wrote of the Chinese that when they wanted to depict an ugly ill-formed man they presented him with short apparel and a large beard, eyes, and nose: Pantoja, *Relación*, 107 ("Letter," in Purchas, *Purchas His Pilgrimes*, 3:367).

22. See, for example, *Kadokawa kogo daijiten*, 2:1.

23. Cited in Berlinguez-Kono, "Debates on *Nichi Zakkyo* in Japan," 20. See also Weiner, "The Invention of Identity," 105–10; Ching, "Yellow Skin, White Masks," 71–75; Iikura, "The Anglo-Japanese Alliance and the Question of Race," 228–29; Sato, "Same Language, Same Race"; and Oguma, *A Genealogy of "Japanese" Self-Images*, 145–47.

24. Iikura, "The 'Yellow Peril' and its Influence on Japanese-German Relations," 81–82; Aydin, *The Politics of Anti-Westernism in Asia*, 56; Askew, "Debating the 'Japanese' Race in Meiji Japan."

25. Henning, "White Mongols?" 155, 161. One school of thought in Russia agreed that Russia was more closely tied to Asia than to Europe; see Mikhailova, "Japan and Russia," 153–54; and Tanaka, *Japan's Orient*, 100–101.

26. For policy debates see Shimazu, *Japan, Race, and Equality*, 95–97. The Buddhist representative's comment was published in "The Religious Bodies in Japan"; see Henning, "White Mongols?" 155; and Clement, *Handbook*, 322–23. *Encyclopaedia Britannica*, 15:275.

27. Kaempfer, *Geschichte und Beschreibung von Japan*, 1:97–117 (*History of Japan*, 1:81–101); Charlevoix, *Histoire et description*, 1:42–43; Siebold, *Nippon*, 1:281–93. On the Ainu see, for example, Baelz, "Menschen-Rassen"; and Koganei, "Beiträge zur physischen Anthropologie der Aino [sic]," esp. 258–61.

28. Griffis, "Are the Japanese Mongolian?"; Knapp, "Who Are the Japanese?"; "Ethnological Basis of the Japanese Claim to Be a White Race"; Mahan, *The Problem of Asia and its Effect on International Policies*, xix; Kowner, "Lighter than Yellow." In modern African American slang a light-skinned black person was also called yellow or "yaller," suggesting another form of "in-betweenness": see *The Greenwood Encyclopedia of African American Slang*, 2:611–12.

29. *Encyclopaedia Britannica*, 6:174, 15:273 n. 1; Richard, *Comprehensive Geography*, 341.

30. *The Book of History*, 1:324, 332, 350. A later chapter on "Qualities of the Japanese People" noted that they had "smooth skins, varying in colour through various yellowish shades, from a hue of brown, in the case of those working in the sun, to a light tint no darker than that of the Southern European" (1:433). On the English of the Far East see also Jones, *Works*, 1:96; and Taylor, *Environment and Race*, 188. On Prussians see Wippich, "Japan-Enthusiasm in Wilhelmine Germany," 74–76.

31. Li, *The Formation of the Chinese People*, 9, 124–25; Liu, "Tentative Classification," 131–32 (and similarly Buxton, *The Peoples of Asia*, 61).

32. Yellow hope: Lehmann, *Image of Japan*, 150; yellow blessing: Sun, "The True Solution of the Chinese Question," in *Guo fu quan ji*, 10:93.

33. Lawrence, *Lectures on Physiology*, 530; Aydin, *Politics of Anti-Westernism*, 54–59, Iikura, "Yellow Peril," 88; on the yellow peril drawing see *Die grosse Politik der europäischen Kabinette*, 9:366; the Japanese scholar is Hishida, *The International Position of Japan as a Great Power*, 262–63.

34. On immigration laws see Ngai, *Impossible Subjects*; for the Supreme Court case see Henning, "White Mongols?" 153; and for Hrdlicka's testimony see *Nonassimilability of Japanese in Hawaii and the United States*, 2, 7–8. On the Chinese as unassimilable see Haller, *Outcasts from Evolution*, 147–52.

35. Petroski, *The Pencil*, 161–63.

36. Myres, "Primitive Man, in Geological Time," 1:22; Prichard, *Researches*, 3rd ed., 1:1–2; Lamprey, "A Contribution to the Ethnology of the

Chinese," 102–3. The presumed yellowness of the Mongolian race had also led Weininger to contend in 1903 that Jews possessed a yellowish skin color because of an admixture of Mongol blood (*Geschlecht und Charakter*, 411).

37. Ross, *The Changing Chinese*, 21; Haddon, *Races of Man*, new ed., 7; Prichard, *Researches*, 3rd ed., 1:217–232. In 1829 Price had used xanthoderm to signify fair skin, not yellow skin (*Essay on the Physiognomy and Physiology of the Present Inhabitants of Britain*, 3–4). One anthropologist offered an interesting twist on yellow skin in 1923: "the natural color [of the Chinese] is very often changed because of the use of opium and morphine, turning the skin to a yellowish and greenish shade" (Shirokogoroff, *Anthropology of Northern China*, 36).

38. Stibbe, *An Introduction to Physical Anthropology*, 149, 152; Hooton, *Up from the Ape*, 548 (and Hooton, *Why Men Behave Like Apes and Vice Versa*, 124); Linton, *The Study of Man*, 42–43; Dixon, *The Racial History of Man*, 281–82, 291; Cruz, *Tractado* 73 (Boxer, *South China*, 137). Macgowan, *Men and Manners of Modern China*, agreed that the Chinese possessed "yellow skin over which no ruddy colour ever passes" (80).

39. Giddings, "The Social Marking System," 725.

40. Montagu, *Statement on Race*, 13, 76 (and remaining unchanged in the third edition of 1972); *What is Race?* 33, 45.

41. Vallois, "L'anthropologie physique," 602, 713, 718; Blumenbach, *De generis humani varietate nativa*, 3rd ed., 120 (*Anthropological Treatises*, 209). In 1977 Bowles noted that while the East Asian group was extremely heterogeneous, their "yellowish tinge" was attributable to "the number and position of pigment granules" (*The People of Asia*, 344, 351).

42. Lye, *America's Asia*, 18; Tsu, *Failure*, 78.

Works Cited

Abel, Clarke. *Narrative of a Journey in the Interior of China.* London, 1818.
Abramson, Marc. "Deep Eyes and High Noses: Physiognomy and the Depiction of Barbarians in Tang China." *Political Frontiers, Ethnic Boundaries, and Human Geographies in Chinese History.* Ed. Nicola di Cosmo and Don J. Wyatt. London: Routledge, 2003. 119–59.
Adachi, Buntaro. "Hautpigment beim Menschen und bei den Affen." *Zeitschrift für Morphologie und Anthropologie* 6 (1903): 1–131.
Adachi, Buntaro, and Kocko Fujisawa. "Mongolen-Kinderfleck bei Europäern." *Zeitschrift für Morphologie und Anthropologie* 6 (1903): 132–33.
Adickes, Erich. *Kant als Naturforscher.* 2 vols. Berlin: de Gruyter, 1924–25.
Aichel, Otto. "Epicanthus, Mongolenfalte, Negerfalte, Hottentottenfalte, Indianerfalte." *Zeitschrift für Morphologie und Anthropologie* 31 (1933): 123–66.
Allen, Gordon, et al. "Mongolism." *The Lancet* (8 April 1961): 775.
Allerhand so lehr- als geist-reiche Brief, Schrifften und Reis-Beschreibungen. Ed. Joseph Stöcklein. 5 vols. Augsburg, 1728–61.
Amiot, Jean Joseph Marie. "L'antiquité des Chinois, prouvée par les monuments." *Mémoires concernant l'histoire, les sciences, les arts, les moeurs, les usages, &c. des Chinois.* 17 vols. Paris, 1776–1814. 2:1–364.
Ammon, Friedrich August von. "Der Epicanthus, ein noch nicht beschriebener gewöhnlich angeborner Fehler des innern Augenwinkels." *Zeitschrift für Ophthalmologie* 1 (1831): 533–39.
———. "Der Epicanthus und das Epiblepharon: zwei Bildungsfehler der menschlichen Gesichtshaut." *Journal für Kinderkrankheiten* 34 (1860): 313–93.
Andersen, Jürgen. *Orientalische Reise-Beschreibungen.* Schleswig, 1669.
André, Jacques. *Étude sur les termes de couleur dans la langue latine.* Paris: Klincksieck, 1949.
Anleitung zu wissenschaftlichen Beobachtungen auf Reisen. Ed. G. Neumayer. Berlin: Oppenheimer, 1875.
Appleton, William W. *A Cycle of Cathay: The Chinese Vogue in England during the Seventeenth and Eighteenth Centuries.* New York: Columbia University Press, 1951.
Ashmead, Albert S. "The Mulberry-Colored Spots on the Skin of the Lower Spine of Japanese and Other Dark Races: A Sign of Negro Descent." *Journal of Cutaneous Diseases Including Syphilis* 23 (1905): 203–14.
———. "Relation of Syphilis with Japanese Racial Peculiarities and Customs." *Atlanta Journal-Record of Medicine* 8 (1906): 395–404.

Askew, David. "Debating the 'Japanese' Race in Meiji Japan: Towards a History of Early Japanese Anthropology." *The Making of Anthropology in East and Southeast Asia*. Ed. Shinji Yamashita et al. New York: Berghahn, 2004. 57–89.

Asnes, Russell S. "Buttock Bruises = Mongolian Spot." *Pediatrics* 74 (1984): 321.

Augstein, H. F. "From the Land of the Bible to the Caucasus and Beyond: The Shifting Ideas of the Geographical Origin of Humankind." *Race, Science, and Medicine, 1700–1960*. Ed. Waltraud Ernst and Bernard Harris. London: Routledge, 1999. 58–79.

Avity, Pierre d'. *The Estates, Empires & Principallities of the World*. London, 1615.

Avril, Philippe. *Travels into Divers Parts of Europe and Asia*. London, 1693.

———. *Voyage en divers états d'Europe et d'Asie*. Paris, 1693.

Aydin, Cemil. *The Politics of Anti-Westernism in Asia: Visions of World Order in Pan-Islamic and Pan-Asian Thought*. New York: Columbia University Press, 2007.

Bachman, John. *The Doctrine of the Unity of the Human Race Examined on the Principles of Science*. Charleston, 1850.

Bacon, Francis. *Silva sylvarum* (1626). *Works*. Ed. James Spedding et al. 14 vols. London: Longman, 1857–74.

Baddeley, John F. *Russia, Mongolia, China*. 2 vols. London: Macmillan, 1919.

Baelz, Erwin. *Awakening Japan: The Diary of a German Doctor*. Trans. Eden Paul and Cedar Paul. New York: Viking 1932.

———. *Erwin Bälz, das Leben eines deutschen Arztes im erwachenden Japan: Tagebücher, Briefe, Berichte*. Stuttgart: Engelhorne, 1930.

———. "Die körperlichen Eigenschaften der Japaner." *Mitteilungen der Deutschen Gesellschaft für Natur- und Völkerkunde Ostasiens* 3 (1880–84): 330–59; 4 (1884–88): 35–103.

———. "Menschen-Rassen Ost-Asiens mit specieller Rücksicht auf Japan." *Verhandlungen der Berliner Gesellschaft für Anthropologie, Ethnologie und Urgeschichte* 33 (1901): 166–89.

———. "Noch einmal die blauen 'Mongolen-Flecke.'" *Internationales Centralblatt für Anthropologie* 7 (1902): 329–31.

———. *Die Ostasiaten: ein Vortrag*. Stuttgart: Wittwer, 1901.

———. "Referat über den Vortrag des Herrn … Dr. E. Baelz … : 'Über die Rassenelemente in Ostasien, speciell in Japan.'" *Mittheilungen der Deutschen Gesellschaft für Natur- und Völkerkunde Ostasiens* 8 (1899–1902): 227–35.

———. "Zur Frage von der Rassen-Verwandtschaft zwischen Mongolen und Indianern." *Verhandlungen der Berliner Gesellschaft für Anthropologie, Ethnologie und Urgeschichte* 33 (1901): 393–94.

Ball, John. *The Modern Practice of Physic; or, a Method of Judiciously Treating the Several Disorders Incident to the Human Body*. 2 vols. London, 1762.

Barbosa, Duarte. *Livro em que dá relação do que viu e ouviu no Oriente*. Lisbon: Agência Geral das Colónias, 1946.

Barkan, Elazar. *The Retreat of Scientific Racism: Changing Concepts of Race in Britain and the United States Between the World Wars*. Cambridge: Cambridge University Press, 1992.
Barnes, Irene. "The Inheritance of Pigmentation in the American Negro." *Human Biology* 1 (1929): 321–81.
Barrow, John. *Travels in China*. London, 1804.
Bartlett, Robert. "Medieval and Modern Concepts of Race and Ethnicity." *Journal of Medieval and Early Modern Studies* 31 (2001): 39–56.
Bartoli, Daniello. *La Cina*. Ed. Bice Garavelli Mortara. Milan: Bompiani, 1975.
———. *Dell'istoria della Compagnia di Gesù: l'Asia* (1653–63). 8 vols. Piacenza, 1819–21.
Basler, Adolf. "Über eine Hautfarbentafel für Europäer und mit vorgenommene Untersuchungen." *Zeitschrift für Morphologie und Anthropologie* 25 (1926): 525–30.
Baur, Erwin, Eugen Fischer, and Fritz Lenz. *Menschliche Erblehre*. 4th ed. Munich: Lehmann, 1936.
———. *Menschliche Erblichkeitslehre*. Munich: Lehmann, 1921.
Beaumont, Saunier de. *Voyage d'Innigo de Biervillas, Portugais*. Paris, 1736.
Beazley, C. Raymond. *The Dawn of Modern Geography*. 3 vols. London: Frowde, 1905–6.
Beddoe, John. "Observations on the Natural Colour of the Skin in Certain Oriental Races." *Journal of the Anthropological Institute of Great Britain and Ireland* 19 (1890): 257–63.
———. *The Races of Britain: A Contribution to the Anthropology of Western Europe*. Bristol: Arrowsmith, 1885.
Belzoni, Giovanni Battista. *Narrative of the Operations and Recent Discoveries within the Pyramids, Temples, Tombs, and Excavations in Egypt and Nubia*. London, 1820.
Benacci, Alessandro. *Avisi venuti novamente da Roma delli XXIII de marzo 1585*. Bologna, 1585.
Benda, Clemens E. *Mongolism and Cretinism*. 2nd ed. New York: Grune & Stratton, 1946.
Bendyshe, Thomas. "The History of Anthropology." *Memoirs Read Before the Anthropological Society of London* 1 (1863–64): 335–458.
Berchet, Guglielmo. "Le antiche ambasciate giapponesi in Italia." *Archivio veneto* 13 (1877): 245–85; 14 (1877): 150–203.
Berlinguez-Kono, Noriko. "Debates on *Nichi Zakkyo* in Japan (1879–99): The Influence of Spencerian Social Evolutionism on the Japanese Perception of the West." *The Japanese and Europe: Images and Perceptions*. Ed. Bert Erdström. Richmond, UK: Japan Library. 7–22.
Bernasconi, Robert. "Who Invented the Concept of Race? Kant's Role in the Enlightenment Construction of Race." *Race*. Ed. Bernasconi. Oxford: Blackwell, 2001. 11–36.

Bernier, François. *Histoire de la dernière révolution des états du Grand Mogol.* 4 vols. Paris, 1670–71.

———. *The History of the Late Revolution of the Empire of the Great Mogol.* London, 1671.

———. "Nouvelle division de la Terre, par les differentes Especes ou Races d'hommes qui l'habitent, envoyée par un fameux Voyageur." *Journal des sçavans* 12 (24 April 1684): 148–55.

Bezzola, Gian Andri. *Die Mongolen in abendländischer Sicht, 1220–1270: ein Beitrag zur Frage der Völkerbegegnungen.* Bern: Francke, 1974.

Bickerton, G. *Accurate Disquisitions in Physick.* London, 1719.

Bindman, David. *Ape to Apollo: Aesthetics and the Idea of Race in the 18th Century.* London: Reaktion, 2002.

Bird, Isabella. *Unbeaten Tracks in Japan: An Account of Travels in the Interior Including Visits to the Aborigines of Yezo and the Shrine of Nikko.* London: Murray, 1880.

Bittner, Stephen, and Eli H. Newberger. "Pediatric Understanding of Child Abuse and Neglect." *Pediatrics in Review* 2 (1981): 197–208.

Blackwood, Beatrice. "Racial Differences in Skin-Colour as Recorded by the Colour Top." *Journal of the Royal Anthropological Institute of Great Britain and Ireland* 60 (1930): 137–68.

Blakeslee, A. F., and D. E. Warner. "Correlation Between Egg-Laying Activity and Yellow Pigment in the Domestic Fowl." *American Naturalist* 49 (1915): 360–68 [also published in *Science* 41 (1915): 432–34].

Bleyer, Adrien. "The Occurrence of Mongolism in Ethiopians." *Journal of the American Medical Association* 84 (1925): 1041–42.

Bloch, Adolphe. "Preuves ataviques de la transformation des races." *Bulletins et mémoires de la Société d'anthropologie de Paris,* 5th series, vol. 2 (1901): 618–24.

Blue, Gregory. "China and Western Social Thought in the Modern Period." *China and Historical Capitalism: Genealogies of Sinological Knowledge.* Ed. Timothy Brook and Gregory Blue. Cambridge: Cambridge University Press, 1999. 57–109.

———. "Gobineau on China: Race Theory, the 'Yellow Peril,' and the Critique of Modernity." *Journal of World History* 10 (1999): 93–139.

Blumenbach, Johann Friedrich. *The Anthropological Treatises of Johann Friedrich Blumenbach.* Trans. Thomas Bendyshe. London: Longman, 1865.

———. *Beyträge zur Naturgeschichte (Erster Theil).* Göttingen, 1790.

———. *Beyträge zur Naturgeschichte (Erster Theil).* 2nd ed. Göttingen, 1806.

———. *Correspondence.* Ed. Frank William Peter Dougherty and Norbert Klatt. 2 vols. Göttingen: Klatt, 2006–7.

———. *De generis humani varietate nativa.* 1st ed. Göttingen, 1775.

———. *De generis humani varietate nativa.* 2nd ed. Göttingen, 1781.

———. *De generis humani varietate nativa.* 3rd ed. Göttingen, 1795.

———. *Handbuch der Naturgeschichte.* Göttingen, 1779.

———. "Observations on Some Egyptian Mummies Opened in London." *Philosophical Transactions of the Royal Society of London* 84 (1794): 177–95.

———. *Über die natürlichen Verschiedenheiten im Menschengeschlechte.* Trans. Johann Gottfried Gruber. Leipzig, 1798.

Bodin, Jean. *The Six Bookes of the Common-Weale.* Trans. Richard Knolles. London, 1606.

———. *Les six livres de la république.* Paris, 1576.

Bodman, G. B. "Color Discs Used in Soil Color Analysis." *Science* 67 (1928): 446–47.

Boemus, Joannes. *Omnium gentium mores, leges et ritus* (1520). Antwerp, 1562.

Bolk, Louis. "On the Origin of Human Races." *Proceedings of the Section of Sciences, Koninklijke Akademie van Wetenschappen, Afdeeling Natuurkunde* 30 (1927): 320–28.

———. "Over mongolenplooi en mongoloide idiotie." *Nederlandsch tijdschrift voor geneeskunde* 67:1 (1923): 226–34.

Bonnett, Alastair. "Who Was White? The Disappearance of Non-European White Identities and the Formation of European Racial Whiteness." *Ethnic and Racial Studies* 21 (1998): 1029–55.

Bontinck, François. *La lutte autour de la liturgie chinoise aux XVIIe et XVIIIe siècles.* Louvain: Nauwelaerts, 1962.

The Book of History: A History of All Nations from the Earliest Times to the Present. 18 vols. New York: Grolier Society, 1915–21.

Booth, Tony. "Labels and Their Consequences." *Current Approaches to Down's Syndrome.* Ed. David Lane and Brian Stratford. London: Cassell, 1985. 3–24.

Bory de Saint-Vincent, Jean Baptiste. *L'homme (homo): essai zoologique sur le genre humain.* 2nd ed. 2 vols. Paris, 1827.

Boscaro, Adriana. "The First Japanese Ambassadors to Europe: Political Background for a Religious Journey." *KBS Bulletin on Japanese Culture* 103 (August–September 1970): 1–20.

———. *Sixteenth-Century European Printed Works on the First Japanese Mission to Europe.* Leiden: Brill, 1973.

Botero, Giovanni. *Relationi universali.* 4 vols. Venice, 1596.

———. *Relations of the Most Famous Kingdomes and Common-wealths Thorowout the World.* London, 1630.

———. *The Worlde; or, An Historicall Description of the Most Famous Kingdomes and Common-weales Therein.* London, 1601.

Bowles, Gordon T. *The People of Asia.* New York: Scribner, 1977.

Bowman, H. A. "The Color-Top Method of Estimating Skin Pigmentation." *American Journal of Physical Anthropology* 14 (1930): 59–72.

Boxer, C. R. *The Christian Century in Japan, 1549–1650.* Berkeley: University of California Press, 1951.

———. "European Missionaries and Chinese Clergy, 1654–1810." *The Age of Partnership: Europeans in Asia Before Dominion.* Ed. Blair B. Kling and M. N. Pearson. Honolulu: University of Hawai'i Press, 1979. 97–121.

———. *South China in the Sixteenth Century: Being the Narratives of Galeote Pereira, Fr. Gaspar da Cruz, O.P., Fr. Martín de Rada, O.E.S.A.* London: Hakluyt Society, 1953.

Bradley, Milton. *Color in the School-Room: A Manual for Teachers.* Springfield, MA: Milton Bradley Co., 1890.

———. "The Color Question Again." *Science* 19 (1892): 175–76.

———. *Elementary Color.* Springfield, MA: Milton Bradley Co., 1895.

Bradley, Richard. *A Philosophical Account of the Works of Nature* (1721). 2nd ed. London, 1739.

Bradley's Kindergarten Material and School Aids. Springfield, MA: Milton Bradley Co., 1904.

Bradshaw, David. "Eugenics: 'They Should Certainly Be Killed.'" *A Concise Companion to Modernism.* Ed. Bradshaw. Malden, MA: Blackwell, 2003. 34–55.

Braude, Benjamin. "The Sons of Noah and the Construction of Ethnic and Geographical Identities in the Medieval and Early Modern Periods." *William and Mary Quarterly* 54 (1997): 103–42.

Brennemann, Joseph. "The Sacral or So-Called 'Mongolian' Pigment Spots of Earliest Infancy and Childhood, with Especial Reference to Their Occurrence in the American Negro." *American Anthropologist* 9 (1907): 12–30.

Broca, Paul. "Échelle chromatique des yeux." *Bulletins de la Société d'anthropologie de Paris* 4 (1863): 592–605.

———. "Instructions générales pour les recherches anthropologiques." *Bulletins de la Société d'anthropologie de Paris* 3 (1862): 411–14.

———. *Instructions générales pour les recherches anthropologiques à faire sur le vivant.* 2nd ed. Paris: Masson, 1879.

———. *Instructions générales pour les recherches et observations anthropologiques (anatomie et physiologie).* Paris: Masson 1865 [also in *Mémoires de la Société d'anthropologie* 2 (1865): 69–204; repr. in *Archives de médecine navale* 3 (1865): 369–504].

———. *Recherches sur l'hybridité animale en général et sur l'hybridité humaine en particulier.* Paris: Claye, 1860.

———. "Tableau chromatique des yeux, de la peau et des cheveux pour les observations anthropologiques." *Bulletins de la Société d'anthropologie de Paris* 5 (1864): 767–73.

Brosses, Charles de. "Essai de géographie étymologique sur les noms donnés aux peuples Scythes anciens & modernes." *Mémoires de l'Académie des sciences, arts et belles lettres de Dijon* 2 (1774): 447–580.

Brousseau, Kate. *Mongolism: A Study of the Physical and Mental Characteristics of Mongolian Imbeciles.* Baltimore: Williams & Wilkins, 1928.

Brown, Judith C. "Courtiers and Christians: The First Japanese Emissaries to Europe." *Renaissance Quarterly* 49 (1994): 872–906.

Brugsch, Henri. *Histoire d'Égypte dès les premiers temps de son existence jusqu'à nos jours.* Leipzig: Hinrich, 1859.

Bruno, Giordano. *Opera latine*. Ed. Francesco Fiorentino. 3 vols. Naples: Morano, 1879–91.

———. *Opere latine*. Ed. Carlo Monti. Turin: Unione Tipografico-Editrice Torinese, 1980.

Buffon, George-Louis Leclerc, comte de. *Histoire naturelle, générale et particulière*. 44 vols. Paris, 1749–1804.

———. *Natural History, General and Particular*. Trans. William Smellie. 2nd ed. 9 vols. London, 1785.

Buxton, L. H. Dudley. *The Peoples of Asia*. New York: Knopf, 1925.

Byard, Pamela J. "Quantitative Genetics of Human Skin Color." *Yearbook of Physical Anthropology* 24 (1981): 123–37.

Cain, A. J. "Linnaeus's Natural and Artificial Arrangements of Plants." *Botanical Journal of the Linnean Society* 117 (1995): 73–133.

Camões, Luis de. *The Lusiads*. Trans. J. J. Aubertin. 2 vols. London: Kegan Paul, 1878.

Camper, Petrus. *Works*. Trans. T. Cogan. London, 1794.

Candidius, George. "Discours ende cort verhael van 't eylant Formosa" (1628). *Archief voor de geschiedenis der oude hollandsche zending*. Ed. J. A. Grothe. 6 vols. Utrecht: Bentum, 1884–91.

Cartas que los padres y hermanos de la Compañia de Jesus, que andan en los reynos de Japon escrivieron a los de la misma Compañia, desde el año de mil y quinientos y quarenta y nueve, hasta el de mil y quinientos y setenta y uno. Alcala, 1575.

Cartas que os padres e irmãos de Companhia de Jesus escreverão dos reynos de Japão & China aos da mesma Companhia da India, & Europa, des do anno de 1549 atè o de 1580. 2 vols. Evora, 1598.

Carus, Carl Gustav. *Denkschrift zum hundertjährigen Geburtsfeste Goethe's: über ungleiche Befähigung der verschiedenen Menschheitstämme für höhere geistige Entwickelung*. Leipzig, 1849.

"Case and Autopsy of a Kalmuc [sic] Idiot." *Journal of Mental Science* 22 (1876): 161–62.

Cecere, Giulia. "Russia and its 'Orient': Ethnographic Exploration of the Russian Empire in the Age of Enlightenment." *The Anthropology of the Enlightenment*. Ed. Larry Wolff and Marco Cipolloni. Stanford: Stanford University Press, 2007. 185–208.

Cecil, Lamar. *Wilhelm II*. 2 vols. Chapel Hill: University of North Carolina Press, 1989–96.

Chambers, Robert. *Vestiges of the Natural History of Creation*. London, 1844.

Chambers, William. *Discours servant d'explication, par Tan Chet-Qua de Quang-Cheou-Fou*. London, 1773.

———. *Treatise of Oriental Gardening*. 2nd ed. London, 1773.

Champollion, Jean-François. *Lettres écrites d'Égypte et de Nubie, en 1828 et 1829*. Paris, 1833.

———. *Monuments de l'Égypte et de la Nubie*. 4 vols. Paris, 1835–45.

Champollion-Figeac, Jacques-Joseph. *Égypte ancienne*. Paris, 1839.

Chardin, John. *Voyages en Perse et autres lieux de l'Orient*. 10 vols. Amsterdam, 1711.

Charlevoix, Pierre-François-Xavier de. *Histoire de l'établissement, des progrès et de la décadence du Christianisme dans l'empire du Japon* (1715). 2 vols. Paris, 1828.

——. *Histoire et description générale du Japon*. 2 vols. Paris, 1736.

Charpentier de Cossigny, Joseph François. *Voyage à Canton*. Paris, 1799.

Chemin, A. "Note sur les taches congénitales de la région sacro-lombaire chez les Annamites." *Bulletins et mémoires de la Société d'anthropologie de Paris*, 4th series, vol. 10 (1899): 130–32.

——. "Taches congénitales de la région sacro-lombaire." *Revue mensuelle de l'École d'anthropologie de Paris* 9 (1899): 196–97.

Chen, William Pai-Dei. *Asian Blepharoplasty and the Eyelid Crease*. 2nd ed. Philadelphia: Elsevier, 2006.

Childs, Donald J. *Modernism and Eugenics: Woolf, Eliot, Yeats, and the Culture of Degeneration*. Cambridge: Cambridge University Press, 2001.

Ching, Leo. "Yellow Skin, White Masks: Race, Class, and Identification in Japanese Colonial Discourse." *Trajectories: Inter-Asia Cultural Studies*. Ed. Kuan-Hsing Chen. London: Routledge, 1998. 65–86.

Chouke, Kehar Singh. "The Epicanthus or Mongolian Fold in Caucasian Children." M.A. thesis. University of Colorado, 1929.

Chow, Kai-Wing. "Imagining Boundaries of Blood: Zhang Binglin and the Invention of the Han 'Race' in Modern China." *The Construction of Racial Identities in China and Japan: Historical and Contemporary Perspectives*. Ed. Frank Dikötter. Honolulu: University of Hawai'i Press, 1997. 34–52.

——. "Narrating Nation, Race, and National Culture: Imagining the Hanzu Identity in Modern China." *Constructing Nationhood in Modern East Asia*. Ed. Chow, Kevin M. Doak, and Poshek Fu. Ann Arbor: University of Michigan Press, 2001. 47–83.

Clement, Ernest W. *A Handbook of Modern Japan*. 9th ed. Chicago: McClurg, 1913.

A Collection of Voyages and Travels. Ed. Awnsham Churchill and John Churchill. 4 vols. London, 1704.

Comas, Juan. *Manual of Physical Anthropology*. Rev. ed. Springfield, IL: Thomas, 1960.

Condorcet, Jean-Antoine-Nicolas de Caritat, marquis de. *Esquisse d'un tableau historique des progrès de l'esprit humain*. Paris, 1795.

——. *Outlines of an Historical View of the Progress of the Human Mind*. London, 1795.

Coon, Carleton S. *The Living Races of Man*. New York: Knopf, 1965.

Cooper, Michael. *The Japanese Mission to Europe, 1582–1590: The Journey of Four Samurai Boys through Portugal, Spain, and Italy*. Folkestone: Global Oriental, 2005.

Cordier, Henri. "L'arrivée des Portugais en Chine." *T'oung pao* 12 (1911): 483–543.

Corrêa, Gaspar. *Lendas da India* (ca. 1550). Ed. Rodrigo José de Lima Felner. 4 vols. Lisbon: Academia Real das Sciencias, 1858–66.

Couto, Diogo do. *Da Asia*. 24 vols. Lisbon, 1777–88.

Crookshank, F. G. *The Mongol in Our Midst: A Study of Man and His Three Faces*. London: Kegan Paul, 1924.

———. *The Mongol in Our Midst: A Study of Man and His Three Faces*. 2nd ed. London: Kegan Paul, 1925.

———. *The Mongol in Our Midst: A Study of Man and His Three Faces*. 3rd ed. London: Kegan Paul, 1931.

———. *Der Mongole in unserer Mitte: ein Studium des Menschen und seiner drei Gesichter*. Trans. Eugen Kurz. Munich: Drei Masken, 1928.

Cruz, Gaspar da. *Tractado em que se contam muito por extenso as cousas da China* (1569). Barcelona: Portucalense, 1937.

Cuvier, Georges. *The Animal Kingdom, Arranged in Conformity with its Organization*. 16 vols. London, 1827–35.

———. *Le règne animal, distribué d'après son organisation*. 4 vols. Paris, 1817.

D'Elia, Pasquale. "Bernardo, il primo giapponese venuto a Roma (1555)." *La civiltà cattolica* 102 (1951): 277–87, 527–35.

Dahlgren, E. W. "A Contribution to the History of the Discovery of Japan." *Transactions and Proceedings of the Japan Society, London* 11 (1914): 239–60.

Dampier, William. *A New Voyage Round the World*. 3 vols. London, 1697–1703.

Daniels, Farrington, Peter W. Post, and Brian E. Johnson. "Theories of the Role of Pigment in the Evolution of Human Races." *Pigmentation: Its Genesis and Biologic Control*. Ed. Vernon Riley. New York: Appleton, 1972. 13–22.

Dante Alighieri. *The Comedy of Dante Alighieri*. Trans. Dorothy L. Sayers. 3 vols. Harmondsworth: Penguin, 1949–62.

———. *La Divina commedia*. Ed. Baldassare Lombardi. 3 vols. Rome, 1791.

———. *Divine Comedy*. Trans. Mark Musa. 6 vols. Bloomington: Indiana University Press, 1996–2004.

———. *Inferno*. Trans. Elio Zappulla. New York: Pantheon, 1998.

Dapper, Olfert [incorrectly ascribed to Arnoldus Montanus]. *Atlas Chinensis*. Trans. John Ogilby. London, 1671.

———. *Gedenkwaerdig bedryf der Nederlandsche Oost-Indische maetschappye, op de kuste en in het keizerrijk van Taising of Sina*. Amsterdam, 1670.

Darwin, Charles. *The Descent of Man*. 2 vols. London: Murray, 1871.

———. *Expression of the Emotions in Man and Animals* (1872). New York: Philosophical Library, 1955.

———. *On the Origin of Species*. London: Murray, 1859.

Davenport, Charles B. *Heredity in Relation to Eugenics*. New York: Holt, 1911.

———. *The Heredity of Skin Color in Negro-White Crosses*. Washington, DC: Carnegie Institution, 1913.

———. "Heredity of Skin Pigment in Man." *American Naturalist* 44 (1910): 641–72, 705–31.

———. *Inheritance in Poultry*. Washington, DC: Carnegie Institution, 1906.

———. "Notes on Physical Anthropology of Australian Aborigines and Black-White Hybrids." *American Journal of Physical Anthropology* 8 (1925): 73–94.

———. "The Skin Colors of the Races of Mankind." *Natural History* 26 (1926): 44–49.

Dawson, Raymond. *The Chinese Chameleon: An Analysis of European Conceptions of Chinese Civilization*. Oxford: Oxford University Press, 1967.

De Moluccis insulis. Cologne, 1523.

Decornoy, Jacques. *Péril jaune, peur blanche*. Paris: Grasset, 1970.

Demel, Walter. "Abundantia, Sapientia, Decadencia: zum Wandel des Chinabildes vom 16. bis zum 18. Jahrhundert." *Die Kenntnis beider "Indien" im frühneuzeitlichen Europa: Akten der Zweiten Sektion des 37. deutschen Historikertages im Bamberg 1988*. Ed. Urs Bitterli and Eberhard Schmitt. Munich: Oldenbourg, 1991. 129–53.

———. *Come i cinesi divennero gialli: alle origini delle teorie razziali*. Milan: Vita e Pensiero, 1997.

———. "The Images of the Japanese and the Chinese in Early Modern Europe: Physical Characteristics, Customs, and Skills: A Comparison of Different Approaches to the Cultures of the Far East." *Itinerario* 25:3–4 (2001): 34–53.

———. "Wie die Chinesen gelb wurden." *Historische Zeitschrift* 255 (1992): 625–66.

Deniker, Joseph. "Étude sur les Kalmouks." *Revue d'anthropologie*, 2nd series, vol. 6 (1883): 671–703; vol. 7 (1884): 277–310, 493–501, 640–79.

———. *Les races et les peuples de la terre* (1900). 2nd ed. Paris: Masson, 1926.

———. *The Races of Man: An Outline of Anthropology and Ethnography*. London: Scott, 1900.

———. "Les taches congénitales dans la région sacro-lombaire considérées comme caractère de race." *Bulletins et mémoires de la Société d'anthropologie de Paris*, 5th series, vol. 2 (1901): 274–81.

Desmoulins, Antoine. *Histoire naturelle des races humaines*. Paris, 1826.

Devisse, Jean, and Michel Mollat. *The Image of the Black in Western Art: From the Early Christian Era to the "Age of Discovery."* 2 vols. Cambridge: Harvard University Press, 1979.

Dikötter, Frank. *The Discourse of Race in Modern China*. Stanford: Stanford University Press, 1992.

———. "Racial Discourse in China: Continuities and Permutations." *The Construction of Racial Identities in China and Japan: Historical and Contemporary Perspectives*. Ed. Dikötter. Honolulu: University of Hawai'i Press, 1997. 12–33.

Diosy, Arthur. *The New Far East*. London: Cassell, 1898.

Dixon, Roland B. *The Racial History of Man*. New York: Scribner, 1923.

Documentação para a história das missões do padroado Português do Oriente: India. Ed. António da Silva Rego. 12 vols. Lisbon: Agência Geral das Colónias, 1947–96.

Documentos del Japón, 1547–1557. Ed. Juan Ruiz de Medina. Rome: Instituto Histórico de la Compañía de Jesús, 1990.

Dorno, Carl. "Beiträge zur Kenntnis des Sonnen- und Quarzlicht-Erythems und -Pigmentes." *Strahlentherapie* 22 (1926): 70–91.

Dougherty, Frank W. P. "Christoph Meiners und Johann Friedrich Blumenbach im Streit um den Begriff der Menschenrasse." *Die Natur des Menschen: Probleme der Physischen Anthropologie und Rassenkunde (1750–1850)*. Ed. Gunter Mann and Franz Dumont. Stuttgart: Fischer, 1990. 89–111.

Down, J. Langdon. "Observations on an Ethnic Classification of Idiots." *Clinical Lectures and Reports by the Medical and Surgical Staff of the London Hospital* 3 (1866): 259–62.

"Down's Sydrome." *The Lancet* (21 October 1961): 935.

Doyle, John P. "Two Sixteenth-Century Jesuits and a Plan to Conquer China." *Rechtsdenken: Schnittpunkte West und Ost: Recht in den gesellschafts- und staatstragenden Institutionen Europas und Chinas*. Ed. Harald Holz and Konrad Wegmann. Munster: LIT, 2005. 253–73.

Drews, Richard. "Ueber das Mongolenauge als provisorische Bildung bei deutschen Kindern und über den Epicanthus." *Archiv für Anthropologie* 18 (1889): 223–33.

Du Halde, Jean-Baptiste. *Description géographique, historique, chronologique, politique, et physique de l'empire de la Chine et de la Tartarie Chinoise*. 4 vols. Paris, 1735.

———. *A Description of the Empire of China*. 2 vols. London, 1738–41.

———. *The General History of China*. 4 vols. London, 1736.

Du Jarric, Pierre. *L'histoire des choses plus mémorables advenuës tant és Indes Orientales*. Valencienne, 1611.

Duméril, André Marie Constant. *Zoologie analytique, ou méthode naturelle de classification des animaux*. Paris, 1806.

Dungy, Claibourne I. "Mongolian Spots, Day Care Centers, and Child Abuse." *Pediatrics* 69 (1982): 672.

Eden, Charles H. *Japan: Historical and Descriptive*. London: Ward, 1877.

Eden, Richard. *De novo orbe; or, the Histories of the West Indies*. London, 1612.

———. *Decades of the Newe Worlde or West India*. London, 1555.

———. *The History of Travayle in the West and East Indies*. London, 1577.

Edwards, Edward A., and S. Quimby Duntley. "The Pigments and Color of Living Human Skin." *American Journal of Anatomy* 65 (1939): 1–33.

Eickstedt, Egon, Freiherr von. *Rassenkunde und Rassengeschichte der Menschheit*. Stuttgart: Enke, 1934.

El Bahrawy, Ali Ahmed. "Über den Mongolenfleck bei Europäern." *Archiv für Dermatologie und Syphilis* 141 (1922): 171–92.

Ellis, Havelock. "The Psychology of Yellow." *Popular Science Monthly* 58 (1906): 456–63.

Encyclopaedia Britannica. 11th ed. 29 vols. Cambridge: Cambridge University Press, 1910–11.

Engel, Johann Jacob. *Der Philosoph für die Welt*. 3 vols. Leipzig, 1775–1800.
Epstein, Alois. "Über den blauen Kreuzfleck und andere mongoloide Erscheinungen bei europäischen Kindern." *Jahrbuch für Kinderheilkunde* 63 (1906): 60–73.
Ernout, A., and A. Meillet. *Dictionnaire étymologique de la langue latine: histoire des mots*. 3rd ed. 2 vols. Paris: Klincksieck, 1951.
Erxleben, Johann Christian Polykarp. *Systema regni animalis*. Leipzig, 1777.
Escalante, Bernardino de. *A Discourse of the Navigation which the Portugales Doe Make to the Realmes and Provinces of the East Partes of the Worlde*. London, 1579.
———. *Discurso de la navegacion que los Portugueses hazen à los Reinos y Provincias del Oriente*. Seville, 1577.
Escalante Alvarado, García de. "Relación del viaje que hizo desde la Neuva-España á las islas del Poniente." *Collección de documentos inéditos, relativos al descubrimiento . . . de las antiguas posesiones españolas* 5 (1866): 117–205.
Eschricht, Daniel Friedrich. *Zoologisch-anatomisch-physiologische Untersuchungen über die nordischen Wallthiere*. Leipzig, 1849.
"Ethnological Basis of the Japanese Claim to Be a White Race." *Current Opinion* 55 (July 1913): 38–39.
Fan, T. C. "Sir William Jones's Chinese Studies." *Review of English Studies* 22 (1946): 304–14.
"The Far Eastern Situation from a German Standpoint." *Review of Reviews* 13 (1896): 3–4.
Farrar, Frederic W. "Aptitudes of Races." *Transactions of the Ethnological Society of London* 5 (1867): 115–26.
Fenves, Peter. "Imagining an Inundation of Australians; or, Leibniz on the Principles of Grace and Race." *Race and Racism in Modern Philosophy*. Ed. Andrew Valls. Ithaca: Cornell University Press, 2005. 73–88.
Ferguson, Donald. "Letters from Portuguese Captives in Canton, Written in 1534 and 1536." *Indian Antiquary* 30 (1901): 421–51, 469–71; 31 (1902): 10–32, 53–65.
Feynes, Henri de. *An Exact and Curious Survey of all the East Indies, Even to Canton, the Chiefe Cittie of China*. London, 1615.
———. *Le voyage de Montferran de Paris a la Chine*. Ed. L.-Marcel Duvic. Paris: Maisonneuve, 1884.
Finlayson, George. *The Mission to Siam and Hué, the Capital of Cochin China, in the Years 1821–2*. London, 1826.
Fischer, Johann Eberhard. *Sibirische Geschichte von der Entdekkung Sibiriens bis aus die Eroberung dieses Lands durch die russische Waffen*. 2 vols. St. Petersburg, 1768.
France, Anatole. *Sur la pierre blanche*. Paris: Calmann-Lévy, 1900.
Franklin, Benjamin. *Papers*. Ed. Leonard W. Labaree et al. 38 vols. New Haven: Yale University Press, 1959–.
Fraser, John, and Arthur Mitchell. "Kalmuc [sic] Idiocy: Report of a Case with Autopsy." *Journal of Mental Science* 22 (1876): 169–79.
Freccero, John. "The Sign of Satan." *MLN* 80 (1965): 11–26.

Friedman, John Block. *The Monstrous Races in Medieval Art and Thought*. Cambridge: Harvard University Press, 1981.
Fritsch, Gustav. "Bemerkungen zu der Hautfarbentafel." *Mitteilungen der Anthropologischen Gesellschaft in Wien* 46 (1916): 183–85.
———. "Die Feststellung der menschlichen Hautfarben." *Zeitschrift für Ethnologie* 48 (1916): 86–89.
Fróis, Luis. *Kulturgegensätze Europa-Japan (1585)*. Ed. Josef Franz Schütte. Tokyo: Sophia University, 1955.
Fryer, John. *A New Account of East-India and Persia*. London, 1698.
Fujisawa, Kocko. "Sogenannter Mongolen-Geburtsfleck der Kreuzhaut bei europäischen Kindern." *Jahrbuch für Kinderheilkunde* 62 (1905): 221–24.
Furukawa, Masashige. "Aesthetic Surgery of Oriental Eyelids." *Aesthetic Plastic Surgery* 1 (1977): 139–43.
Galvão, António. *The Discoveries of the World, from Their First Original unto the Year of Our Lord 1555* (1601). Ed. C. R. Drinkwater Bethune. London: Hakluyt Society, 1862.
García, Gregorio. *Origen de los Indios de el nuevo mundo e Indias Occidentales* (1607). 2nd ed. Madrid, 1729.
Gates, R. Ruggles. *Pedigrees of Negro Families*. Philadelphia: Blakiston, 1949.
———. "Studies of Interracial Crossing: Spectrophotometric Measurements of Skin Color." *Human Biology* 24 (1952): 25–34.
Gaubil, Antoine. *Histoire de Gentchiscan et de toute la dinastie des Mongous*. Paris, 1739.
———. *Traité de la chronologie Chinoise*. Ed. A. I. Silvestre de Sacy. *Mémoires concernant l'histoire, les sciences, les arts, les moeurs, les usages, &c. des Chinois*. 17 vols. Paris, 1776–1814. Vol. 17.
Gemelli Careri, Giovanni Francesco. *Giro del mondo*. 6 vols. Naples, 1699–1700.
A General History of the Turks, Moguls, and Tatars, Vulgarly called Tartars. 2 vols. London, 1729–30.
Geoffroy Saint-Hilaire, Isidore. "Sur la classification anthropologique et particulièrement sur les types principaux du genre humain." *Mémoires de la Société d'anthropologie de Paris* 1 (1860–63): 125–44.
Georgi, Johann Gottlieb. *Bemerkungen einer Reise im russischen Reich im Jahre 1772–1774*. 2 vols. St. Petersburg, 1775.
———. *Beschreibung aller Nationen des russischen Reichs*. St. Petersburg, 1776–80.
———. *Russia; or, a Compleat Account of All the Nations Which Compose That Empire*. 4 vols. London, 1780–83.
"German Sympathies." *The Times* (9 September 1904): 4.
Germann, Susanne. *Ein Leben in Ostasien: die unveröffentlichten Reisetagebücher des Arztes, Anthropologen und Ethnologen Erwin Baelz (1849–1913)*. Bietigheim-Bissingen: Archiv der Stadt Bietigheim-Bissingen, 2006.
Gibbon, Edward. *History of the Decline and Fall of the Roman Empire* (1776–88). 3 vols. New York: Modern Library, 1952–55.

Gibson, David. *Down's Syndrome: The Psychology of Mongolism.* Cambridge University Press, 1978.
Giddings, Franklin H. "The Social Marking System." *American Journal of Sociology* 15 (1910): 721–40.
Girtanner, Christoph. *Über das kantische Prinzip für die Naturgeschichte.* Göttingen, 1796.
Gobineau, Arthur, comte de. *Essai sur l'inégalité des races humaines.* 4 vols. Paris: Didot, 1853–55.
———. *The Moral and Intellectual Diversity of Races.* Philadelphia: Lippincott, 1856.
———. *Oeuvres.* Ed. Jean Gaulmier. 3 vols. Paris: Gallimard, 1983–87.
Goethe, Johann Wolfgang von. *Goethe's Color Theory.* Ed. Rupprecht Matthaei. New York: Van Nostrand Reinhold, 1971.
———. *Werke: Hamburger Ausgabe.* Ed. Erich Trunz. 15 vols. Hamburg: Wegner, 1962–65.
Goldenberg, David M. *The Curse of Ham: Race and Slavery in Early Judaism, Christianity, and Islam.* Princeton: Princeton University Press, 2003.
Goldsmith, Oliver. *An History of the Earth, and Animated Nature.* 8 vols. London, 1774.
Gollwitzer, Heinz. *Die gelbe Gefahr: Geschichte eines Schlagworts, Studien zum imperialistischen Denken.* Göttingen: Vandenhoeck & Ruprecht, 1962.
González de Mendoza, Juan. *See* Mendoza, Juan González de.
Gould, Stephen Jay. *The Mismeasure of Man.* 2nd ed. New York: Norton, 1996.
———. *The Panda's Thumb: More Reflections in Natural History.* New York: Norton, 1980.
Graberg da Hemsö, Jacopo. "Lettera di Giovanni da Empoli a Leonardo suo padre intorno al viaggio da lui fatto a Malacca e frammenti di altre lettere." *Archivio storico italiano*, Appendix 3 (1846): 35–91.
Graf, Arturo. *Miti, leggende e superstitioni del medio evo.* 2 vols. Turin: Loescher, 1892–93 [repr. New York: Burt Franklin, 1971].
The Greenwood Encyclopedia of African American Slang. Ed. Anand Prahlad. 3 vols. Westport: Greenwood, 2006.
Griffis, William Elliot. "Are the Japanese Mongolian?" *North American Review* 197 (January–June 1913): 721–33.
Grosier, Jean-Baptiste. *Description générale de la Chine.* 2 vols. Paris, 1787.
———. *A General Description of China.* 2 vols. London, 1788.
Die grosse Politik der europäischen Kabinette, 1871–1914: Sammlung der diplomatischen Akten des Auswärtigen Amtes. 40 vols. Ed. Johannes Lepsius et al. Berlin: Deutsche Verlagsgesellschaft für Politik und Geschichte, 1922–27.
Gualtieri, Guido. *Relationi della venuta degli ambasciatori giaponesi à Roma.* Rome, 1586.
Guignes, Chrétien-Louis-Joseph de. *Voyages à Peking, Manille et l'Île de France.* 3 vols. Paris, 1808.

Guignes, Joseph de. *Histoire générale des Huns, des Turcs, des Mogols, et des autres Tartares occidentaux.* 4 vols. Paris, 1756–58.

Gulick, Sidney L. *The White Peril in the Far East: An Interpretation of the Significance of the Russo-Japanese War.* New York: Revell, 1905.

Gumilla, Joseph. *El Orinoco illustrado* (1741). 2nd ed. 2 vols. Madrid, 1745.

Guthrie, R. Dale. "The Mammoth Steppe and the Origin of Mongoloids and Their Dispersal." *Prehistoric Mongoloid Dispersals.* Ed. Takeru Akazawa and Emöke J. E. Szathmary. Oxford: Oxford University Press, 1995. 172–86.

Gutierrez, Beniamino. *La prima ambascieria giapponese in Italia: dall'ignorata cronaca di un diarista e cosmografo milanese della fine del XVI sec.* Milan: Perego, 1938.

Gützlaff, Karl Friedrich August. *A Sketch of Chinese History, Ancient and Modern.* 2 vols. New York, 1834.

Guzmán, Luis de. *Historia de las misiones de la Compañía de Jesús en la India oriental* (1601). Bilbao: El Mensajero del corazon de Jesús, 1891.

Haddon, A. C. *The Races of Man and Their Distribution.* London: Milner, 1909.

———. *The Races of Man and Their Distribution.* New ed. Cambridge: Cambridge University Press, 1924.

———. *The Study of Man.* London: Bliss, 1898.

Hahn, Thomas. "The Difference the Middle Ages Makes: Color and Race Before the Modern World." *Journal of Medieval and Early Modern Studies* 31 (2001): 1–37.

Hakluyt, Richard. *The Principal Navigations, Voyages, Traffiques and Discoveries of the English Nation.* 3 vols. London, 1599–1600.

Hall, Jonathan M. *Ethnic Identity in Greek Antiquity.* Cambridge: Cambridge University Press, 1997.

Haller, John S. *Outcasts from Evolution: Scientific Attitudes of Racial Inferiority, 1859–1900.* Urbana: University of Illinois Press, 1971.

Handwörterbuch des deutschen Aberglaubens. Ed. E. Hoffmann-Krayer. 10 vols. Berlin: de Gruyter, 1927–42.

Hansen, Soren. "Bidrag til Østgrønlaendernes Anthropologi." *Meddelelser om Grønland* 10 (1888): 1–41.

———. "Bidrag til Vestgrønlaendernes Anthropologi." *Meddelelser om Grønland* 7 (1893): 163–248.

Hardy, Arthur C. "A New Recording Spectrophotometer." *Journal of the Optical Society of America* 25 (1935): 305–19.

Harman, P. M. *The Natural Philosophy of James Clerk Maxwell.* Cambridge: Cambridge University Press, 1998.

Harris, J. Arthur, A. F. Blakeslee, and D. E. Warner. "The Correlation Between Body Pigmentation and Egg Production in the Domestic Fowl." *Genetics* 2 (1917): 36–77.

Harris, Marvin. *The Rise of Anthropological Theory: A History of Theories of Culture.* New York: Crowell, 1968.

Harris, Reginald G. "The San Blas Indians." *American Journal of Physical Anthropology* 9 (1926): 17–63.

Hatch, J. Leffingwell. "Some Studies upon the Chinese Brain." *Internationale Monatsschrift für Anatomie und Physiologie* 8 (1891): 101–10.
Hay, John. *De rebus japonicis, indicis, et peruanis epistolae recentiores*. Antwerp, 1605.
Hazart, Cornelius. *Kerckelycke historie van de gheheele wereldt*. 4 vols. Antwerp, 1668–82.
———. *Kirchen-Geschichte*. 2nd ed. Vienna, 1694.
Heinrich, Larissa N. *The Afterlife of Images: Translating the Pathological Body Between China and the West*. Durham: Duke University Press, 2008.
Henning, Joseph M. "White Mongols?: The War and American Discourses on Race and Religion." *The Impact of the Russo-Japanese War*. Ed. Rotem Kowner. London: Routledge, 2007. 153–66.
Herder, Johann Gottfried. *Outlines of a Philosophy of the History of Man*. Trans. T. Churchill. London, 1800.
———. *Sämtliche Werke*. Ed. Bernhard Suphan. 33 vols. Berlin: Weidmann, 1877–1913 [repr. Hildesheim: Olms, 1994].
Herrman, Charles. "Mongolian Imbecility as an Anthropologic Problem." *Archives of Pediatrics* 42 (1925): 523–29.
Herskovits, Melville J. *The American Negro: A Study in Racial Crossing*. New York: Knopf, 1928.
———. *The Anthropometry of the American Negro*. New York: Columbia University Press, 1930.
———. "Does the Negro Know His Father?" *Opportunity* 4 (1926): 306–10.
Heylyn, Peter. *Cosmographie*. 2nd ed. London, 1657.
———. *Microcosmus*. London, 1621.
Hintze, Arthur. "Der Hautfarbenfächer und das Hautfarbendiagramm." *Zeitschrift für Ethnologie* 59 (1927): 254–78.
Hishida, Seiji G. *The International Position of Japan as a Great Power*. New York: Columbia University Press, 1905.
Histoire généalogique des Tatars. 2 vols. Leiden, 1726.
Hodgen, Margaret T. *Early Anthropology in the Sixteenth and Seventeenth Centuries*. Philadelphia: University of Pennsylvania Press, 1964.
Hooton, Earnest Albert. *Up from the Ape*. New York: Macmillan, 1931.
———. *Why Men Behave Like Apes and Vice Versa; or, Body and Behavior*. Princeton: Princeton University Press, 1941.
Horn, Georg. *Arca Noae*. Leiden, 1666.
———. *De originibus americanis*. The Hague, 1652.
Hornung, Erik. *Ägyptische Unterweltsbücher*. 2nd ed. Zurich: Artemis, 1984.
———. *The Ancient Egyptian Books of the Afterlife*. Trans. David Lorton. Ithaca: Cornell University Press, 1999.
———. *The Tomb of Pharaoh Seti I / Das Grab Sethos' I*. Zurich: Artemis, 1991.
———. *The Valley of the Kings: Horizon of Eternity*. Trans. David Warburton. New York: Timken, 1990.

Howard-Jones, Norman. "On the Diagnostic Term 'Down's Disease.'" *Medical History* 23 (1979): 102–4.
Howell, James. *Lexicon tetraglotton*. London, 1660.
Hrdlicka, Ales. "Anthropometry." *American Journal of Physical Anthropology* 2 (1919): 43–67, 175–94, 283–319, 401–28; 3 (1920): 147–73 [repr. as *Anthropometry*. Philadelphia: Wistar Institute of Anatomy and Biology, 1920].
———. *Directions for Collecting Information and Specimens for Physical Anthropology*. Washington, DC: Government Printing Office, 1904.
———. *Hrdlicka's Practical Anthropometry*. 4th ed. Ed. T. D. Stewart. Philadelphia: Wistar Institute of Anatomy and Biology, 1952.
———. *The Old Americans*. Baltimore: Williams & Wilkins, 1925.
Hsia, Adrian. *Deutsche Denker über China*. Frankfurt am Main: Insel, 1985.
Huard, Pierre. *Chinese Medicine*. Trans. Bernard Fielding. New York: McGraw-Hill, 1968.
———. "Depuis quand avons-nous la notion d'une race jaune?" *Bulletins et travaux, Institut indochinois pour l'étude de l'homme* 4 (1941): 41–42.
Huddleston, Lee Eldridge. *Origins of the American Indians: European Concepts, 1492–1729*. Austin: University of Texas Press, 1967.
Hudson, Nicholas. "From 'Nation' to 'Race': The Origin of Racial Classification in Eighteenth-Century Thought." *Eighteenth-Century Studies* 29 (1996): 247–64.
Hulme, Peter. "Black, Yellow, and White on St. Vincent: Moreau de Jonnès's Carib Ethnography." *The Global Eighteenth Century*. Ed. Felicity A. Nussbaum. Baltimore: Johns Hopkins University Press, 2003. 182–94.
Humboldt, Alexander von. *Voyage aux régions équinoxiales du nouveau continent fait en 1799, 1800, 1801, 1802, 1803 et 1804*. 3 vols. Paris, 1814–34.
Hume, David. *Essays, Moral and Political*. 3rd ed. London, 1748.
[Hunt, James.] "Race in Legislation and Political Economy." *Anthropological Review* 4 (1866): 113–35.
Huxley, Thomas H. "On the Distribution of the Races of Mankind, and its Bearing on the Antiquity of Man." *International Congress of Prehistoric Archaeology: Transactions of the Third Session*. London: Longman, 1869. 92–105.
———. "On the Geographical Distribution of the Chief Modifications of Mankind." *Journal of the Ethnological Society of London* 2 (1870): 404–12.
Iikura, Akira. "The Anglo-Japanese Alliance and the Question of Race." *The Anglo-Japanese Alliance, 1902–1922*. Ed. Phillips Payson O'Brien. London: Routledge, 2004. 222–35.
———. "The 'Yellow Peril' and its Influence on Japanese-German Relations." *Japanese-German Relations, 1895–1945: War, Diplomacy, and Public Opinion*. Ed. Christian W. Sprang and Rolf-Harald Wippich. London: Routledge, 2006. 80–97.
Ijzerman, J. W. *Dirck Gerritsz Pomp, alias Dirck Gerritsz China: de eerste Nederlander die China en Japan bezocht (1544–1604)*. The Hague: Nijhoff, 1915.

"Instruction générale adressée aux voyageurs." *Mémoires de la Société ethnologique* 1 (1841): vi–xv [repr. in *Journal of the Asiatic Society of Bengal* 10 (1841): 175–81].
Instructions générales aux voyageurs. Paris: Delagrave, 1875.
Isaac, Benjamin. *The Invention of Racism in Classical Antiquity.* Princeton: Princeton University Press, 2004.
Isidore of Seville. *Etymologiae. Patrologiae cursus completus. Series Latina.* Ed. J.-P. Migne. 221 vols. Paris, 1844–64. Vol. 82.
———. *Etymologies.* Ed. Stephen A. Barney et al. Cambridge: Cambridge University Press, 2006.
Ives, Edward. *Reisen nach Indien und Persien.* Trans. Christian Wilhelm von Dohm. 2 vols. Leipzig, 1774–75.
———. *Voyage from England to India in the Year 1754.* London, 1773.
Iwanowski, Alexis. "Zur Anthropologie der Mongolen." *Archiv für Anthropologie* 24 (1897): 65–90.
Jackson, John P., and Nadine M. Weidman. *Race, Racism, and Science: Social Impact and Interaction.* Santa Barbara: ABC-CLIO, 2004.
Jackson, Mark. "Changing Depictions of Disease: Race, Representation, and the History of 'Mongolism.'" *Race, Science, and Medicine, 1700–1960.* Ed. Waltraud Ernst and Bernard Harris. London: Routledge, 1999. 167–88.
Japan in Europa: Texte und Bilddokumente zur europäischen Japankenntnis von Marco Polo bis Wilhelm von Humboldt. Ed. Peter Kapitza. 3 vols. Munich: Iudicium, 1990.
Jones, Andrew F. *Yellow Music: Media Culture and Colonial Modernity in the Chinese Jazz Age.* Durham: Duke University Press, 2001.
Jones, David Martin. *The Image of China in Western Social and Political Thought.* Basingstoke: Palgrave, 2001.
Jones, Sir William. *The Letters of Sir William Jones.* Ed. Garland Cannon. 2 vols. Oxford: Clarendon Press, 1970.
———. "On the Second Classical Book of the Chinese." *Asiatic Researches; or, Transactions of the Society Instituted in Bengal* 2 (1799): 195–204.
———. *Works.* 6 vols. London, 1799.
Junker, Thomas. "Blumenbach's Racial Geometry." *Isis* 89 (1998): 498–501.
Kadokawa kogo daijiten [Kadokawa Dictionary of Ancient Japanese]. 5 vols. Tokyo: Kadokawa, 1982–99.
Kaempfer, Engelbert. *De beschryving van Japan.* The Hague, 1729.
———. *Geschichte und Beschreibung von Japan.* Ed. Christian Wilhelm Dohm. 2 vols. Lemgo, 1777–79.
———. *Histoire naturelle, civile, et ecclésiastique de l'empire du Japon.* 2 vols. The Hague, 1729.
———. *The History of Japan.* 2 vols. London, 1727.
Kammerer, Albert. "La découverte de la Chine par les Portugais au XVIème siècle et la cartographie des Portulans." *T'oung pao* 39 (Supplement 1944): 1–260.

Kant, Immanuel. "Bestimmung des Begriffs einer Menschenrace." *Berlinische Monatsschrift* 6 (1785): 390–417.
———. *Kant's Werke: Akademie Textausgabe*. 9 vols. Berlin: de Gruyter, 1910–23.
———. *Von den verschiedenen Racen der Menschen*. Königsberg, 1775.
Kao, Aloys. "Essai sur l'antiquité des Chinois." *Mémoires concernant l'histoire, les sciences, les arts, les moeurs, les usages, &c. des Chinois*. 17 vols. Paris, 1776–1814. 1:1–271.
Keevak, Michael. *The Pretended Asian: George Psalmanazar's Eighteenth-Century Formosan Hoax*. Detroit: Wayne State University Press, 2004.
Keith, Arthur. *Human Embryology and Morphology*. 3rd ed. London: Arnold, 1913.
———. "A New Theory of the Descent of Man." *Nature* (15 December 1910): 206.
Kevles, Daniel J. *In the Name of Eugenics: Genetics and the Uses of Human Heredity*. New York: Knopf, 1985.
Klaatsch, Hermann. "Die Aurignac-Rasse und ihre Stellung im Stammbaum der Menschheit." *Zeitschrift für Ethnologie* 42 (1910): 513–77.
———. *The Evolution and Progress of Mankind*. Trans. Joseph McCabe. New York: Stokes, 1923.
———. "Menschenrassen und Menschenaffen." *Korrespondenz-Blatt der Deutschen Gesellschaft für Anthropologie, Ethnologie und Urgeschichte* 41 (1910): 91–100.
———. *Der Werdegang der Menschheit und die Entstehung der Kultur*. Berlin: Bong, 1920.
Klaatsch, Hermann, and Otto Hauser. "Homo Aurignacensis Hauseri, ein paläolithischer Skeletfund aus dem unteren Aurignacien der Station Combe-Capelle bei Montferrand (Périgord)." *Prähistorische Zeitschrift* 1 (1910): 273–338.
Klaproth, Julius von. *Mémoires relatifs à l'Asie*. 3 vols. Paris, 1826–28.
Klemm, Gustav. *Allgemeine Cultur-Geschichte der Menschheit*. 10 vols. Leipzig, 1843–52.
Klopprogge, Axel. *Ursprung und Ausprägung des abendländischen Mongolenbildes im 13. Jahrhundert: ein Versuch zur Ideengeschichte des Mittelalters*. Wiesbaden: Harrassowitz, 1993.
Knapp, Arthur May. "Who Are the Japanese?" *Atlantic Monthly* 110 (September 1912): 333–40.
Koerner, Lisbet. *Linnaeus: Nature and Nation*. Cambridge: Harvard University Press, 1999.
Koganei, Yoshikiyo. "Beiträge zur physischen Anthropologie der Aino [sic]." *Mittheilungen aus der Medicinische Facultät der Kaiserlich-Japanischen Universität* 2 (1894): 1–404.
Kornerup, A., and J. H. Wanscher. *Methuen Handbook of Colour*. London: Methuen, 1963.
Kowner, Rotem. "'Lighter Than Yellow, but Not Enough': Western Discourse on the Japanese 'Race,' 1854–1904." *Historical Journal* 43 (2000): 103–31.
———. "Skin as a Metaphor: Early European Racial Views on Japan, 1548–1853." *Ethnohistory* 51 (2004): 751–78.

Kupperman, Karen Ordahl. *Indians and English: Facing Off in Early America*. Ithaca: Cornell University Press, 2000.

Kurz, Eugen. "Das Chinesengehirn." *Zeitschrift für Anatomie und Entwicklungsgeschichte* 72 (1924): 199–382.

———. "Ergebnisse rassenanatomischer Untersuchungen über das Extremitätenskelet des Chinesen." *Archiv für Anatomie und Physiologie: Anatomische Abteilung* 43 (1919): 185–203.

———. "Der Unterkiefer des Chinesen." *Archiv für Anatomie und Physiologie: Anatomische Abteilung* 42 (1918): 173–209.

———. "Untersuchungen zur Anatomie der Weichteile beim Chinesen unter Berücksichtigung des Verhaltens bei den Affen." *Zeitschrift für Anatomie und Entwicklungsgeschichte* 67 (1923): 232–85.

———. "Zwei Chinesengehirne: ein Beitrag zur Rassenanatomie." *Zeitschrift für Morphologie und Anthropologie* 16 (1913–14): 281–328.

Labat, Jean-Baptiste. *Nouvelle relation de l'afrique occidentale*. 5 vols. Paris, 1728.

Lach, Donald F. "China in Western Thought and Culture." *Dictionary of the History of Ideas*. Ed. Philip P. Wiener. 5 vols. New York: Scribner, 1973–74. 1:353–73.

Lach, Donald F., and Edwin J. Van Kley. *Asia in the Making of Europe*. 3 vols. Chicago: University of Chicago Press, 1965–93.

Laffey, John F. *Imperialism and Ideology: An Historical Perspective*. Montreal: Black Rose, 2000.

Lamprey, J. "A Contribution to the Ethnology of the Chinese." *Transactions of the Ethnological Society of London* 6 (1868): 101–8.

Lanci, Michelangelo. *Paralipomeni all'illustrazione della sagra scrittura*. 2 vols. Paris, 1845.

Langdon-Down, Reginald. "Some Observations on the Mongolian Type of Imbecility." *Journal of Mental Science* 52 (1906): 187–90.

Larsen, Nils Paul, and Lois Stewart Godfrey. "Sacral Pigment Spots: A Record of Seven Hundred Cases with a Genetic Theory to Explain its Occurrence." *American Journal of Physical Anthropology* 10 (1927): 253–74.

Latham, R. G. *The Natural History of the Varieties of Man*. London, 1850.

Lawrence, James B. *China and Japan and a Voyage Thither: An Account of a Cruise in the Waters of the East Indies, China, and Japan*. Hartford: Case, Lockwood & Brainard, 1870.

Lawrence, William. *Lectures on Physiology, Zoology, and the Natural History of Man*. London, 1819.

———. *A Treatise on the Diseases of the Eye*. London, 1833.

Le Comte, Louis. *Memoirs and Observations . . . Made in a Late Journey through the Empire of China*. London, 1698.

———. *Nouveaux mémoires sur l'état présent de la Chine*. 2 vols. Paris, 1696.

Le Dantec, Aristide. *Précis de pathologie exotique*. 3rd ed. 2 vols. Paris: Doin, 1911.

Lefébure, Eugène. "Les races connues des Égyptiens." *Annales du Musée Guimet* 1 (1880): 61–76.

Legendre, Aimé François. *La civilisation chinoise moderne.* Paris: Payot, 1926.
Lehmann, Jean-Pierre. *The Image of Japan: From Feudal Isolation to World Power, 1850-1905.* London: Allen & Unwin, 1978.
Leibniz, Gottfried Wilhelm. *Otium hanoveranum.* Leipzig, 1718.
——. *Sämtliche Schriften und Briefe, erste Reihe: allgemeiner politischer und historischer Briefwechsel.* 20 vols. Darmstadt: Reichl, 1923-.
Lepsius, C. R. *Denkmaeler aus Aegypten und Aethiopien.* 12 vols. Berlin, 1849-56.
——. *Denkmäler aus Aegypten und Aethiopien, Text.* Ed. Eduard Naville et al. 5 vols. Leipzig: Hinrich, 1897-1913.
Lettres édifiantes et curieuses, écrites des missions étrangères par quelques missionaires de la Compagnie de Jésus. 34 vols. Paris, 1703-76.
Letts, Malcolm. "Prester John: Sources and Illustrations." *Notes and Queries* 188 (January-June 1945): 178-80, 204-7, 246-48, 266-68; 189 (July-December 1945): 4-7.
Lewis, Martin W., and Kären E. Wigen. *The Myth of Continents: A Critique of Metageography.* Berkeley: University of California Press, 1997.
Li, Chi. *The Formation of the Chinese People: An Anthropological Inquiry.* Cambridge: Harvard University Press, 1928.
Lichtenberg, Georg Christoph. *Vermischte Schriften.* Ed. Ludwig Christian Lichtenberg and Friedrich Kries. 9 vols. Göttingen, 1800-1806.
Linnaeus, Carl. *The Animal Kingdom, or Zoological System, of the Celebrated Sir Charles Linnaeus.* Trans. Robert Kerr. London, 1792.
——. *Clavis medicinae.* Stockholm, 1766.
——. *Fundamenta botanica.* Amsterdam, 1736.
——. *A General System of Nature.* Trans. William Turton. 7 vols. London, 1802-6.
——. *Philosophia botanica.* Stockholm, 1751.
——. *Systema naturae.* 1st ed. Leiden, 1735.
——. *Systema naturae.* 2nd ed. Stockholm, 1740.
——. *Systema naturae.* 3rd ed. Halle, 1740.
——. *Systema naturae.* 6th ed. Stockholm, 1748.
——. *Systema naturae.* 10th ed. 2 vols. Stockholm, 1758-59.
——. *Systema naturae.* 12th ed. 3 vols. Stockholm, 1766-68.
——. *Systema naturae.* 13th ed. Ed. Johann Friedrich Gmelin. 3 vols. Leipzig, 1788-93.
——. *Vollständiges Natursystem.* Trans. Philipp Ludwig Statius Miller. 6 vols. Nuremberg, 1773-75.
Linschoten, Jan Huyghen van. *John Huighen van Linschoten His Discours of Voyages into ye Easte & West Indies.* London, 1598.
Linton, Ralph. *The Study of Man: An Introduction.* New York: Appleton, 1936.
Liu, Chungshee H. "A Tentative Classification of the Races of China." *Zeitschrift für Rassenkunde* 6 (1937): 129-50.
Liu, Don, and Wen Ming Hsu. "Oriental Eyelids: Anatomic Difference and Surgical Consideration." *Ophthalmic Plastic and Reconstructive Surgery* 2 (1986): 59-64.

Livro que trata das cousas da India e do Japão. Ed. Adelino de Almeida Calado. Coimbra: Universidade de Coimbra, 1957.

Loarca, Miguel de. "Relación del viaje que hezimos a la China desde la ciudad de Manila en 1575." http://www.upf.edu/asia/projectes/che/s16/loarca.htm.

Lopes de Castanheda, Fernão. *História do descobrimento & conquista da India pelos Portugueses* (1553). 3rd ed. Ed. Pedro de Azevedo. 4 vols. Coimbra: Universidade de Coimbra, 1924–33.

López de Gómara, Francisco. *Histoire generalle des Indes Occidentales.* Paris, 1569.

———. *Historia general de las Indias.* Zaragoza, 1555.

Loureiro, Rui Manuel. *Fidalgos, Missionários e Mandarins: Portugal e a China no século XVI.* Lisbon: Fundação Oriente, 2000.

Lovejoy, Arthur O. *The Great Chain of Being: A Study of the History of an Idea.* Cambridge: Harvard University Press, 1936.

Lowe, Ronald F. "The Eyes in Mongolism." *British Journal of Ophthalmology* 33 (1949): 131–74.

Lucena, João de. *História da vida do padre Francisco de Xavier.* Lisbon, 1600.

Luschan, Felix von. "Einige wesentliche Fortschritte in der Technik der physischen Anthropologie." *Zeitschrift für Ethnologie* 36 (1904): 465–66.

———. "Über Hautfarbentafeln." *Zeitschrift für Ethnologie* 48 (1916): 402–6.

Lye, Colleen. *America's Asia: Racial Form and American Literature, 1893–1945.* Princeton: Princeton University Press, 2005.

Lyman, Stanford. "The 'Yellow Peril' Mystique." *International Journal of Politics, Culture, and Society* 13 (2006): 683–747.

MacFarlane, Charles. *Japan: An Account, Geographical and Historical* (1852). Hartford: Andrus, 1865.

Macgowan, J. *Men and Manners of Modern China.* London: Unwin, 1912.

Maffei, Giovanni Pietro. *L'histoire des Indes Orientales* (1604). Paris, 1665.

———. *Historiarum indicarum* (1588). Bergamo, 1590.

———. *Le istorie dell'Indie Orientali.* Venice, 1589.

Mahan, A. T. *The Problem of Asia and its Effect on International Policies.* Boston: Little, Brown, 1900.

Mailla, Joseph de. *Histoire générale de la Chine, ou annales de cet empire.* 13 vols. Paris, 1777–85.

Malte-Brun, Conrad. *Précis de la géographie universelle.* 8 vols. Paris, 1810–29.

———. *Universal Geography.* 6 vols. Philadelphia, 1827–32.

Malthus, T. R. *An Essay on the Principle of Population, as it Affects the Future Improvement of Society.* London, 1798.

Mandelslo, Johann Albrecht von. *Morgenländischen Reyse-Beschreibung.* Schleswig, 1658.

Mandeville, John. *Mandeville's Travels: Texts and Translations.* Ed. Malcolm Letts. 2 vols. London: Hakluyt Society, 1953.

Markus, Andrew. *Fear and Hatred: Purifying Australia and California, 1850–1901.* Sydney: Hale & Iremonger, 1979.

Marsy, François-Marie de. *Histoire moderne des Chinois, des Japonois, des Indiens, des Persans, des Turcs, des Russes, etc.* 30 vols. Paris, 1755–71.

Martin, Rudolf. *Lehrbuch der Anthropologie in systematischer Darstellung.* 2nd ed. 3 vols. Jena: Fischer, 1928.

Martini, Martino. *Novus atlas sinensis.* Amsterdam, 1655.

Massarella, Derek. "Envoys and Illusions: The Japanese Embassy to Europe, 1582–90: *De missione legatorum japonensium* and the Portuguese Viceregal Embassy to Toyotomi Hideyoshi, 1591." *Journal of the Royal Asiatic Society* 15 (2005): 329–50.

Matignon, J. J. "Stigmates congénitaux et transitoires chez les Chinois." *Bulletins et mémoires de la Société d'anthropologie de Paris,* 4th series, vol. 7 (1896): 524–38.

Maxwell, James Clerk. *Scientific Papers.* Ed. W. D. Niven. 2 vols. Cambridge: Cambridge University Press, 1890.

Mayerhofer, E. "Gegen die Mongolentheorie der sog. 'Mongolenflecke.'" *Zeitschrift für die gesamte experimentelle Medizin* 60 (1928): 255–70.

Mayes, Stanley. *The Great Belzoni.* New York: Putnam, 1959.

McClellan, Robert. *The Heathen Chinee: A Study of American Attitudes toward China, 1890–1905.* Columbus: Ohio State University Press, 1971.

McLaughlin, Peter. "Kant on Heredity and Adaptation." *Heredity Produced: At the Crossroads of Biology, Politics, and Culture, 1500–1870.* Ed. Staffan Müller-Wille and Hans-Jörg Rheinberger. Cambridge: MIT Press, 2007. 277–91.

Mehnert, Ute. *Deutschland, Amerika und die "gelbe Gefahr": zur Karriere eines Schlagworts in der großen Politik, 1905–1917.* Stuttgart: Steiner, 1995.

Meijer, Miriam Claude. *Race and Aesthetics in the Anthropology of Petrus Camper (1722–1789).* Amsterdam: Rodolpi, 1999.

Meiners, Christoph. *Grundriß der Geschichte der Menschheit.* Lemgo, 1785.

———. *Grundriß der Geschichte der Menschheit.* 2nd ed. Lemgo, 1793.

Mellin, G.S.A. *Encyclopädisches Wörterbuch der kritischen Philosophie.* 6 vols. Jena, 1797–1804.

Mendoza, Juan González de. *Histoire du grand royaume de la Chine.* Paris, 1589.

———. *Historia de las cosas mas notables, ritos y costumbres del gran reyno de la China.* Rome, 1585.

———. *L'historia del gran regno della China.* Venice, 1587.

———. *The Historie of the Great and Mightie Kingdome of China.* London, 1588.

———. *De historie ofte beschrijvinghe van het groote rijck van China.* Amsterdam, 1595.

———. *The History of the Great and Mighty Kingdom of China.* Ed. George T. Staunton. 2 vols. London: Hakluyt Society, 1853–54.

———. *Ein neuwe, kurze, doch warhafftige Beschreibung dess . . . Königreichs China.* Frankfurt am Main, 1589.

———. *Nova et succincta, vera tamen historia de . . . regno China.* Frankfurt am Main, 1589.

———. *Rerum morumque in regno Chinensi.* Antwerp, 1655.

Meriwether, C. "A Sketch of the Life of Date Masamune and an Account of His Embassy to Rome." *Transactions of the Asiatic Society of Japan* 21 (1893): 1–105.
Merton, Thomas A. *Mankind in the Unmaking: The Anthropology of Mongolism.* Fairlight, Australia: Thomas A. Merton, 1968.
Metchnikov, Elie [as Elias Metschnikoff]. "Ueber die Beschaffenheit der Augenlider bei den Mongolen und Kaukasiern." *Zeitschrift für Ethnologie* 6 (1874): 153–60.
Meyer, Eduard. "Bericht über eine Expedition nach Ägypten zur Erforschung der Darstellungen der Fremdvölker." *Sitzungsberichte der Preussischen Akademie der Wissenschaften* 1913: 769–801.
Mikhailova, Yulia. "Japan and Russia: Mutual Images, 1904–1939." *The Japanese and Europe: Images and Perceptions.* Ed. Bert Erdström. Richmond, UK: Japan Library. 152–71.
Miller, Konrad. *Mappaemundi: die ältesten Weltkarten.* 6 vols. Stuttgart: Roth, 1895–98.
Miller, Stuart Creighton. *The Unwelcome Immigrant: The American Image of the Chinese, 1785–1882.* Berkeley: University of California Press, 1969.
Minutoli, Johann Heinrich Carl, Freiherr von. *Reise zum Tempel des Jupiter Ammon in der libyschen Wüste, und nach Ober-Aegypten in den Jahren 1820 und 1821.* Berlin, 1827.
Mitchell, John. "An Essay upon the Causes of the Different Colours of People in Different Climates." *Philosophical Transactions of the Royal Society of London* 43 (1744–45): 102–50.
Moges, marquis de. *Souvenirs d'une ambassade en Chine et au Japon en 1857 et 1858.* Paris: Hachette, 1860.
Mongolism. Ed. G.E.W. Wolstenholme and Ruth Porter. Boston: Little, Brown, 1967.
Montagu, Ashley. *Statement on Race: An Extended Discussion in Plain Language of the UNESCO Statement by Experts on Race Problems.* New York: Schuman, 1951.
Montalboddo, Francanzano da. *Paesi novamente retrovati & novo mondo da Alberico Vesputio Florentino intitulato.* Milan, 1508.
Montanus, Arnoldus. *Ambassades mémorables de la Compagnie des Indes Orientales des Provinces Unies vers les empereurs du Japon.* Amsterdam, 1680.
———. *Atlas Chinensis.* See Dapper, Olfert.
———. *Atlas Japannensis.* Trans. John Ogilby. London, 1670.
———. *Gedenkwaerdige gesantschappen der Oost-Indische Maatschappy in 't Vereenigde Nederland, aan de kaisaren van Japan.* Amsterdam, 1669.
Montesquieu, Charles de Secondat, baron de. *Oeuvres complètes.* Ed. Roger Callois. 2 vols. Paris: Gallimard, 1949.
———. *Oeuvres complètes.* Ed. André Masson. 3 vols. Paris: Nagel, 1950–55.
Moran, J. F. *The Japanese and the Jesuits: Alessandro Valignano in Sixteenth-Century Japan.* London: Routledge, 1993.
Morton, Samuel George. *Crania Aegyptiaca; or, Observations on Egyptian Ethnography Derived from Anatomy, History, and the Monuments.* Philadelphia, 1844.
———. *Crania Americana; or, a Comparative View of the Skulls of Various Aboriginal Nations of North and South America.* Philadelphia, 1839.

———. "Observations on the Size of the Brain in Various Races and Families of Man." *Proceedings of the Academy of Natural Sciences Philadelphia* 4 (1848–49): 221–24.
Müller, Friedrich. *Allgemeine Ethnographie*. Vienna: Hölder, 1873.
Mungello, David E. *Curious Land: Jesuit Accommodation and the Origins of Sinology*. Honolulu: University of Hawai'i Press, 1989.
———. *The Great Encounter of China and the West, 1500–1800*. 3rd ed. Lanham, MD: Rowman & Littlefield, 2009.
Münster, Sebastian. *Cosmographia*. Basel, 1552.
———. *A Treatyse of the Newe India, with Other New Founde Landes and Ilandes*. Trans. Richard Eden. London, 1553.
Myers, Henry A. "Marvels, Myths, and Misconceptions: The Classical World and the Far East." *Western Views of China and the Far East*. Ed. Myers. 2 vols. Hong Kong: Asian Research Service, 1982–84. 1:43–83.
Myres, John L. "Primitive Man, in Geological Time." *The Cambridge Ancient History*. Ed. J. B. Bury et al. 12 vols. Cambridge: Cambridge University Press, 1923–29. 1:1–56.
Nakamura, Hirosi. "Passage en France de Hasekura, Ambassadeur Japonais à la cour de Rome au commencement du XVIIe siècle." *Monumenta Nipponica* 3 (1940): 445–57.
Needham, Joseph. *Science and Civilisation in China*. 7 vols. Cambridge: Cambridge University Press, 1954–.
A New Collection of Voyages and Travels. Ed. Thomas Astley. 4 vols. London, 1745–47.
A New Collection of Voyages, Discoveries, and Travels. 7 vols. London, 1767.
Ngai, Mae M. *Impossible Subjects: Illegal Aliens and the Making of Modern America*. Princeton: Princeton University Press, 2004.
Nichols, John. *Illustrations of the Literary History of the Eighteenth Century*. 8 vols. London, 1817–58.
Nicolle de la Croix, Louis Antoine. *Géographie moderne, précédée d'un petit traité de la sphère & du globe* (1752). 2 vols. Paris, 1766.
Nieuhof, Johannes. *Het gezantschap der Neerlandtsche Oost-Indische Compagnie, aan den grooten Tartarischen Cham, den tegenwoordigen keizer van China*. Amsterdam, 1665.
Nonassimilability of Japanese in Hawaii and the United States: Hearings Before the Committee on the Territories, House of Representatives, Sixty-Seventh Congress, Second Session. Washington, DC: Government Printing Office, 1922.
Noort, Olivier van. *De reis om de wereld* (1602). Ed. J. W. Ijzerman. 2 vols. The Hague: Nijhoff, 1926.
Notes and Queries on Anthropology. 2nd ed. London: Anthropological Institute, 1892.
Notes and Queries on Anthropology for the Use of Travellers and Residents in Uncivilized Lands. London: Stanford, 1874.
Nott, Josiah C., and George R. Gliddon. *Types of Mankind; or, Ethnological Researches, Based upon the Ancient Monuments, Paintings, Sculptures, and Crania of Races*. Philadelphia: Lippincott, 1854.

Novus orbis regionum ac insularum veteribus incognitarum. Basel, 1532.
Oguma, Eiji. *A Genealogy of "Japanese" Self-Images*. Trans. David Askew. Melbourne: Trans Pacific, 2002.
Oken, Lorenz. *Allgemeine Naturgeschichte für alle Stände*. 7 vols. Stuttgart, 1833–42.
Ong, Seng. "Wang-y-Tong." *Old Sennockian Newsletter* (Easter 2006): 17.
Onishi, Katsutomo. "Honpojin no kenretsu (ganken haretsu)" [The palpebral fissure in Japanese people]. *Nippon ganka gakkai zasshi* [Japanese Journal of Ophthalmology] 3 (1899): 397–432, 452–552, 585–640.
The Origins of Modern Psychiatry. Ed. Chris Thompson. Chichester: Wiley, 1987.
Ortelius, Abraham. *Theatrum orbis terrarum*. Antwerp, 1587.
Osbeck, Pehr. *Dagbok öfwer en ostindisk resa åren 1750, 1751, 1752*. Stockholm, 1757.
———. *Reise nach Ostindien und China*. Rostock, 1765.
———. *A Voyage to China and the East Indies*. 2 vols. London, 1771.
Osborne, Roy. "Telesio's Dictionary of Latin Color Terms." *Color Research and Application* 27 (2002): 140–46.
Osório, Jerónimo. *De rebus Emmanuelis* (1571). 3 vols. Coimbra, 1791.
———. *The History of the Portuguese during the Reign of Emmanuel*. 2 vols. London, 1752.
Palafox y Mendoza, Juan de. *The History of the Conquest of China by the Tartars*. London, 1671.
Pallas, Peter Simon. *Reise durch verschiedene Provinzen des russischen Reichs*. 3 vols. St. Petersburg, 1771–76.
———. *Sammlungen historischer Nachrichten über die mongolischen Völkerschaften*. 2 vols. St. Petersburg, 1776–1801.
Palmer, Leroy S., and Harry L. Kempster. "The Influence of Specific Feeds and Certain Pigments on the Color of the Egg Yolk and Body Fat of Fowls." *Journal of Biological Chemistry* 39 (1919): 331–37.
Pan, Lynn. *Sons of the Yellow Emperor: A History of the Chinese Diaspora*. Boston: Little, Brown, 1990.
Pantoja, Diego de. *Relación de la entrada de algunos padres de la Compañia de Jesús en la China*. Valencia, 1606.
Parker, Peter. *Journal of an Expedition from Singapore to Japan*. London, 1838.
Parry, William Edward. *Journal of a Second Voyage for the Discovery of a North-West Passage from the Atlantic to the Pacific*. London, 1824.
Paske-Smith, M. *Western Barbarians in Japan and Formosa in Tokugawa Days, 1603–1868*. Kobe: Thompson, 1930.
Pastoureau, Michel. *Couleurs, images, symboles: études d'histoire et d'anthropologie*. Paris: Léopard d'Or, 1989.
———. *Figures et couleurs: études sur la symbolique et la sensibilité médiévales*. Paris: Léopard d'Or, 1986.
———. "Formes et couleurs du désordre: le jaune avec le vert." *Médiévales* 4 (1983): 62–73.

Pauly, August Friedrich von. *Paulys Realencyclopädie der classischen Altertumswissenschaft.* New ed. Ed. Georg Wissowa. 49 vols. Stuttgart: Metzler, 1893–1978.
Pauthier, Guillaume. *Chine, ou description historique, géographique et littéraire de ce vaste empire, d'après des documents chinois.* Paris, 1837.
Pearce, Susan M. "Giovanni Battista Belzoni's Exhibition of the Reconstructed Tomb of Pharoah Seti I in 1821." *Journal of the History of Collections* 12 (2000): 109–25.
Pearson, Karl. "Note on the Skin-Colour of the Crosses Between Negro and White." *Biometrika* 6 (1908–9): 348–53.
Penrose, Lionel. "The Blood Grouping of Mongolian Imbeciles." *The Lancet* (20 February 1932): 394–95.
Pétits de la Croix, François. *Histoire du grand Genghizcan premier empereur des anciens Mogols et Tartares.* Paris, 1710.
———. *The History of Genghizcan the Great, First Emperor of the Antient Moguls and Tartars.* London, 1722.
Petroski, Henry. *The Pencil: A History of Design and Circumstance.* New York: Knopf, 1990.
"A Picture by Emperor William." *Harper's Weekly* (22 January 1898): 76.
Pigafetta, Antonio. *Le voyage et navigation faict par les Espaignols es Isles de Mollucques.* Paris, 1525.
Pillsbury, J. H. "A New Color Scheme." *Science* 19 (1892): 114.
Pinkerton, John. *Modern Geography* (1802). 2 vols. Philadelphia, 1804.
Pires, A. Thomas. "O Japão no seculo XVI." *O Instituto* 54 (1907): 54–64.
Pires, Tomé. *Suma Oriental.* Trans. Armando Cortesão. 2 vols. London: Hakluyt Society, 1944.
Pöch, Rudolf. "Zweiter Bericht über die von der Wiener Anthropologischen Gesellschaft in den k.u.k. Kriegsgefangenenlagern veranlaßten Studien." *Mitteilungen der Anthropologischen Gesellschaft in Wien* 46 (1916): 107–31.
Polo, Marco. *The Book of Ser Marco Polo the Venetian, Concerning the Kingdoms and Marvels of the East.* Trans. Henry Yule. 3rd ed. 2 vols. New York: Scribner, 1926.
Poole, Reginald Stuart. "The Egyptian Classification of the Races of Man." *Journal of the Anthropological Institute of Great Britain and Ireland* 16 (1887): 370–79.
Pownall, Thomas. *The Administration of the Colonies.* 2nd ed. London, 1765.
Prévost, Abbé. *Allgemeine Historie der Reisen zu Wasser und Lande.* 21 vols. Leipzig, 1747–74.
———. *Histoire général des voyages.* 19 vols. Paris, 1746–89.
Price, Thomas. *Essay on the Physiognomy and Physiology of the Present Inhabitants of Britain.* London, 1829.
Prichard, James Cowles. *The Natural History of Man.* 2nd ed. London, 1845.
———. *Researches into the Physical History of Man.* London, 1813.
———. *Researches into the Physical History of Man.* 3rd ed. 5 vols. London, 1836–47.
Purchas, Samuel. *Purchas His Pilgrimes.* 4 vols. London, 1625.

Pusey, James Reeve. *China and Charles Darwin.* Cambridge: Harvard University Press, 1983.
Quatrefages, Armand de. *L'espèce humaine.* Paris: Ballière, 1877.
———. *Rapport sur les progrès de l'anthropologie.* Paris: Imprimerie impériale, 1867.
Race and the Enlightenment: A Reader. Ed. Emmanuel Chukwudi Eze. Oxford: Blackwell, 1997.
Rachewiltz, I. de. *Prester John and Europe's Discovery of East Asia.* Canberra: Australian National University Press, 1972.
Rada, Martín de. *Relaçion verdadera delascosas del reyno de Taibin por otro nombre China* (1575). http://www.upf.edu/asia/projectes/che/s16/radapar.htm.
Radulet, Carmen M. *Vasco da Gama: la prima circumnavigazione dell'Africa, 1497–1499.* Reggio Emilia: Diabasis, 1994.
Raffles, Stamford. *The History of Java.* 2 vols. London, 1817.
———. *Report on Japan to the Secret Committee of the English East India Company.* Kobe: Thompson, 1929.
Rafter, Nicole. *The Criminal Brain: Understanding Biological Theories of Crime.* New York: New York University Press, 2008.
Ramusio, Giovanni Battista. *Delle navigationi et viaggi.* 2nd ed. 3 vols. Venice, 1554–59.
———. *Navigazioni e viaggi.* Ed. Marica Milanesi. 6 vols. Turin: Einaudi, 1978–88.
Ranke, Johannes. "Ueber das Mongolenauge als provisorische Bildung bei deutschen Kindern." *Korrespondenz-Blatt der Deutschen Gesellschaft für Anthropologie, Ethnologie und Urgeschichte* 19 (1888): 115–18.
Ranke, Karl Ernst. "Über die Hautfarbe der südamerikanischen Indianer." *Zeitschrift für Ethnologie* 30 (1898): 61–73.
Ratsimamanga, A. Rakoto. "Tache pigmentaire héréditaire et origines des Malgaches." *Revue anthropologique* 50 (1940): 5–128.
Ratzel, Friedrich. *Die chinesische Auswanderung: ein Betrag zur Cultur- und Handelsgeographie.* Breslau: Kern, 1876.
Receuil de voyages au nord. 3rd ed. 10 vols. Amsterdam, 1731–38.
Reitz, M. "Die Hautfarbe der alten Ägypter." *Kleine Kulturgeschichte der Haut.* Ed. Ernst G. Jung. Darmstadt: Steinkopff, 2007. 156–65.
Relation de la Grande Tartarie dressée sur les mémoires originaux des suèdois prisonniers en Sibérie pendant la guerre de la Suède avec la Russie. Amsterdam, 1737.
Relatione del viaggio et arrivo in Europa et Roma de' principi giapponesi. Venice, 1585.
"The Religious Bodies in Japan." *Japan Weekly Mail* (21 May 1904): 580.
Rémusat, Abel. *Nouveaux mélanges asiatiques.* 2 vols. Paris, 1829.
———. *Recherches sur les langues tartares.* Paris, 1820.
Review of Antoine Desmoulins, *Tableaux analytiques de la zoologie. Bulletin des sciences naturelles et de géologie* 6 (1825): 239–45.
Ricci, Matteo. *China in the Sixteenth Century: The Journals of Matthew Ricci, 1583–1610.* Trans. Louis J. Gallagher. New York: Random House, 1953.

———. *De Christiana expeditione apud Sinas suscepta ab Societate Jesu*. Ed. Nicolas Trigault. Augsburg, 1615.

———. *Entrata nella China de' padri della Compagnia del Gesù (1582–1610)* (1622). Trans. Antonio Sozzini. Rome: Paoline, 1983.

———. *Fonti Ricciane*. Ed. Pasquale M. D'Elia. 3 vols. Rome: Libreria dello Stato, 1942–49.

———. *Histoire de l'expedition chrestienne au royaume de la Chine entreprinse par les peres de la Compagnie de Jesus*. Lille, 1617.

———. *Istoria de la China i cristiana empresa hecha en ella por la Compañia de Jesus*. Seville, 1621.

Richard, Louis. *Comprehensive Geography of the Chinese Empire*. Trans. M. Kennelly. Shanghai: T'usewei, 1908.

———. *Géographie de l'empire de Chine (cours supérieur)*. Shanghai: Mission catholique, 1905.

Rivet, Paul. *Les origines de l'homme américain*. 2nd ed. Paris: Gallimard, 1957.

Robins, Ashley H. *Biological Perspectives on Human Pigmentation*. Cambridge: Cambridge University Press, 1991.

Rogaski, Ruth. *Hygienic Modernity: Meanings of Health and Disease in Treaty-Port China*. Berkeley: University of California Press, 2004.

Rogers, Spencer L. *The Colors of Mankind: The Range and Role of Human Pigmentation*. Springfield, IL: Thomas, 1990.

Röhl, John C. G. *Wilhelm II: der Aufbau der Persönlichen Monarchie, 1888–1900*. Munich: Beck, 2001.

Rosellini, Ippolito. *I monumenti dell'Egitto e della Nubia*. 9 vols. Pisa, 1832–44.

Ross, E. Denison. "Prester John and the Empire of Ethiopia." *Travel and Travellers of the Middle Ages*. Ed. Arthur Percival Newton. London: Kegan Paul, 1926. 174–94.

Ross, Edward Alsworth. *The Changing Chinese: The Conflict of Oriental and Western Cultures in China*. London: Unwin, 1911.

Rossiianov, Kirill. "Taming the Primitive: Elie Metchnikov and His Discovery of Immune Cells." *Osiris* 23 (2008): 213–29.

Rowntree, Leonard G., and George E. Brown. "A Tintometer for the Analysis of the Color of the Skin." *American Journal of the Medical Sciences* 170 (1925): 341–47.

Rudolphi, Karl Asmund. *Grundriss der Physiologie*. 2 vols. Berlin, 1821–28.

Rule, Paul A. *K'ung-tzu or Confucius? The Jesuit Interpretation of Confucianism*. Sydney: Allen & Unwin, 1986.

Saabye, Hans Egede. *Greenland: Being Extracts from a Journal Kept in That Country in the Years 1770 to 1778*. London, 1818.

Saint-Pierre, Bernardin de. *Études de la nature* (1784). 2nd ed. 4 vols. Paris, 1787–88.

———. *Studies of Nature*. 5 vols. London, 1796.

Sánchez, Alonso. "De la entrada de la China en particular." *Labor evangelica: ministerios apostolicos de los obreros de la Compañía de Jesús, fundacion, y progressos *de su provincia en las islas Filipinas*. New ed. Ed. Pablo Pastells. 3 vols. Barcelona: Henrich, 1900–1902. 1:438–45.

———. "The Proposed Entry into China, in Detail." *The Philippine Islands, 1493–1898*. Ed. Emma Helen Blair and James Alexander Robertson. 55 vols. Cleveland: Clark, 1903–9. 6:197–230.

Sande, Duarte de. *De missione legatorum japonensium ad Romanam curiam*. Macao, 1590.

Sapir, Edward. "The Race Problem." *The Nation* (1 July 1925): 40–42.

Sarasin, Paul, and Fritz Sarasin. *Ergebnisse naturwissenschaftlicher Forschungen auf Ceylon*. 4 vols. Wiesbaden: Kreidel, 1887–1908.

———. *Materialien zur Naturgeschichte der Insel Celebes*. 5 vols. Wiesbaden: Kreidel, 1898–1906.

Saris, John. *The Voyage of Captain John Saris to Japan, 1613*. Ed. Ernest M. Satow. London: Hakluyt Society, 1900.

Sato, Kazuki. "'Same Language, Same Race': The Dilemma of *Kanbun* in Modern Japan." *The Anglo-Japanese Alliance, 1902–1922*. Ed. Phillips Payson O'Brien. London: Routledge, 2004. 118–35.

Saunders, J. J. "Matthew Paris and the Mongols." *Essays in Medieval History Presented to Bertie Wilkinson*. Ed. T. A. Sandquist and M. R. Powicke. Toronto: University of Toronto Press, 1969. 116–32.

Schiller, Francis. *Paul Broca: Founder of French Anthropology, Explorer of the Brain*. Berkeley: University of California Press, 1979.

Schlapp, Max G. "Mongolism: A Chemical Phenomenon." *Journal of Heredity* 16 (1925): 161–70.

Schön, J.M.A. *Handbuch der pathologischen Anatomie des menschlichen Auges*. Hamburg, 1828.

———. "Zur Geschichte des Epicanthus." *Zeitschrift für Ophthalmologie* 2 (1832): 120–22.

Schurhammer, Georg. "O descobrimento do Japão pelos Portugueses no ano de 1543." *Orientalia*. Lisbon: Centro de Estudos Históricos Ultramarinos, 1963. 485–580.

———. "Der 'Grosse Brief' des Heiligen Franz Xaver." *Xaveriana*. Lisbon: Centro de Estudos Históricos Ultramarinos, 1964. 605–29.

Schütte, Josef Franz. *Valignano's Mission Principles for Japan*. Trans. John J. Coyne. 2 vols. St. Louis: Institute of Jesuit Sources, 1980–85.

Semedo, Alvaro. *Relatione della grande monarchia della Cina*. Rome, 1643.

Sena, Tereza. "Macau: o primeiro ponto de encontro permanente na China." *Revista de cultura: RC* 27–28 (April–September 1996): 25–59.

Sera, G. L. "I caratteri della faccia e il polifiletismo dei primati." *Giornale per la morfologia dell'uomo e dei primati* 2 (1918): 1–296.

———. "Sul significato polifiletico delle differenze strutturali nell'arto inferiore di Anthropoidea." *Giornale per la morfologia dell'uomo e dei primati* 3 (1919–21): 85–181.

Serres, Etienne. "Principes d'embryogénie, de zoogénie et de tératogénie." *Mémoires de l'Académie des sciences de l'Institut de France* 25 (1860): 1-942.
Serruys, Henry. "Mongol *Altan* 'Gold' = 'Imperial.'" *Monumenta Serica* 21 (1962): 357-78.
Shah, Nayan. *Contagious Divides: Epidemics and Race in San Francisco's Chinatown*. Berkeley: University of California Press, 2001.
Shaxby, J. H., and H. E. Bonnell. "On Skin Colour." *Man* 28 (1928): 60-64.
Sheard, Charles, and George E. Brown. "The Spectrophotometric Analysis of the Color of the Skin." *Archives of Internal Medicine* 38 (1926): 816-31.
Sheard, Charles, and Louis A. Brunsting. "The Color of the Skin as Analyzed by Spectrophotometric Methods." *Journal of Clinical Investigation* 7 (1929): 559-613.
Sherman, Paul D. *Colour Vision in the Nineteenth Century*. Bristol: Hilger, 1981.
Shimazu, Naoko. *Japan, Race, and Equality: The Racial Equity Proposal of 1919*. London: Routledge, 1998.
Shirokogoroff, S. M. *Anthropology of Northern China*. Shanghai: Royal Asiatic Society, North China Branch, 1923.
Shoemaker, Nancy. "How Indians Got to Be Red." *American Historical Review* 102 (1997): 625-44.
———. *A Strange Likeness: Becoming Red and White in Eighteenth-Century North America*. Oxford: Oxford University Press, 2004.
Shuttleworth, G. E. "Mongolian Imbecility." *British Medical Journal* (11 September 1909): 661-65.
———. "The Physical Features of Idiocy, in Relation to Classification and Prognosis." *Liverpool Medico-Chirurgical Journal* 3 (1883): 283-301.
Sichel, Jules. "Mémoire sur l'épicanthus et sur une espèce particulière et non encore décrite de tumeur lacrymale." *Annales d'oculistique* 26 (1851): 29-58.
Siebold, Philipp Franz von. *Nippon: Archiv zur Beschreibung von Japan* (1832). 2nd ed. 2 vols. Würzburg: Woerl, 1897.
Singleton, Charles S. "Commedia, Elements of Structure." *Dante Studies* 1 (1954): 33-42.
Sinica franciscana. Ed. Anastasius van den Wyngaert. 11 vols. Florence: Collegium S. Bonaventurae, 1929-.
Slessarev, Vsevolod. *Prester John: The Letter and the Legend*. Minneapolis: University of Minnesota Press, 1959.
Slotkin, J. S. *Readings in Early Anthropology*. New York: Wenner-Gren Foundation for Anthropological Research, 1965.
Smedley, Audrey. *Race in North America: Origin and Evolution of a Worldview*. Boulder: Westview, 1993.
Smialek, John E. "Significance of Mongolian Spots." *Journal of Pediatrics* 97 (1980): 504.
Smith, Charles Hamilton. *The Natural History of the Human Species*. Boston: Gould and Lincoln, 1851.

Smith, Samuel Stanhope. *An Essay on the Causes of the Variety of Complexion and Figure in the Human Species.* 2nd ed. New Brunswick, 1810.

Snowden, Frank M. *Before Color Prejudice: The Ancient View of Blacks.* Cambridge: Harvard University Press, 1983.

Solinus, C. Julius. *Collectanea rerum memorabilium.* Ed. Theodor Mommsen. Berlin: Weidmann, 1895.

Sollors, Werner. *Neither Black Nor White Yet Both: Thematic Explorations of Interracial Literature.* Oxford: Oxford University Press, 1997.

Solórzano Pereira, Juan de. *Política indiana.* Madrid, 1648.

A Source Book in Medieval Science. Ed. Edward Grant. Cambridge: Harvard University Press, 1974.

Standaert, Nicolas. *The Interweaving of Rituals: Funerals in the Cultural Exchange Between China and Europe.* Seattle: University of Washington Press, 2008.

Staunton, George. *An Authentic Account of an Embassy from the King of Great Britain to the Emperor of China.* 2 vols. London, 1797.

Stepan, Nancy. *The Idea of Race in Science: Great Britain, 1800–1960.* London: Macmillan, 1982.

Stevens, John. *A New Spanish and English Dictionary.* London, 1706.

Stibbe, E. P. *An Introduction to Physical Anthropology.* London: Arnold, 1930.

Strahlenberg, Philipp Johann von. *An Histori-Geographical Description of the North and Eastern Part of Europe and Asia.* London, 1730.

———. *Das nord- und ostliche Theil von Europa und Asia.* Stockholm, 1730.

Strong, R. M. "A Quantitative Study of Variation in the Smaller North-American Shrikes." *American Naturalist* 35 (1901): 271–98.

Struys, Jan Janszoon. *Drie aanmerkelyke reizen* (1676). Amsterdam, 1746.

Stuurman, Siep. "François Bernier and the Invention of Racial Classification." *History Workshop Journal* 50 (2000): 1–21.

Sullivan, Louis R. *Essentials of Anthropometry: A Handbook for Explorers and Museum Collectors.* New York: American Museum of Natural History, 1928.

Sumner, Francis B. "Linear and Colorimetric Measurements of Small Mammals." *Journal of Mammalogy* 8 (1927): 177–206.

Sun Yatsen. *Guo fu quan ji* [The Complete Works of Sun Yat-sen]. 12 vols. Taipei: Jindai Zhongguo chubanshe, 1989. 10:87–96.

———. *San min chu i: The Three Principles of the People.* Trans. Frank W. Price. Shanghai: China Committee, Institute of Pacific Relations, 1927.

Tanaka, Stefan. *Japan's Orient: Rendering Pasts into History.* Berkeley: University of California Press, 1993.

Tavernier, Jean-Baptiste. *Six voyages de Jean-Baptiste Tavernier, écuyer baron d'Aubonne, qu'il a fait en Turquie, en Perse, et aux Indes.* 2 vols. Paris, 1676–77.

———. *The Six Voyages of John Baptista Tavernier, Baron of Aubonne, through Turky, into Persia and the East-Indies.* 2 vols. London, 1677.

Taylor, Griffith. *Environment and Race: A Study of the Evolution, Migration, Settlement, and Status of the Races of Man.* Oxford: Oxford University Press, 1927.

Telesio, Antonio. *Opera*. Naples, 1762.
ten Kate, Herman F. "Die blauen Geburtsflecke." *Globus* 87 (1905): 53–58.
———. "Neue Mitteilungen über die blauen Geburtsflecken." *Zeitschrift für Ethnologie* 37 (1905): 756–58.
The Texts and Versions of John de Plano Carpini and William de Rubruquis. Ed. C. Raymond Beazley. London: Hakluyt Society, 1903.
Thomas, Franklin. *The Environmental Basis of Society: A Study in the History of Sociological Theory*. New York: Century, 1925.
Thompson, Richard Austin. *The Yellow Peril, 1890–1924* (1957). New York: Arno Press, 1978.
Thomson, Arthur. "On the Treatment and Utilization of Anthropological Data." *Knowledge* (1 February 1899): 25–27.
Thomson, Mathew. *Psychological Subjects: Identity, Culture, and Health in Twentieth-Century Britain*. Oxford: Oxford University Press, 2006.
Thunberg, Carl Peter. *Inträdes-tal, om de mynt-sorter, som i äldre och sednare tider blifvit slagne och varit gångbare uti Kejsaredömet Japan*. Stockholm, 1779.
———. "Ett kort utdrag af en journal, hållen på en resa til och uti Keisaredömet Japan." *Philosophical Transactions of the Royal Society of London* 70 (1780): 143–56.
———. *Reise durch einen Theil von Europa, Afrika und Asien, hauptsächlich in Japan, in den Jahren 1770 bis 1779*. 2 vols. Berlin, 1794.
———. *Resa uti Europa, Africa, Asia, förrättad åren 1770–1779*. 4 vols. Uppsala, 1788–93.
———. *Travels in Europe, Africa, and Asia, Performed Between the Years 1770 and 1779*. 2nd. ed. 4 vols. London, 1795.
———. *Verhandeling over de Japansche natie*. Amsterdam, 1780.
———. *Voyage en Afrique et en Asie, principalement au Japon, pendant les années 1770–1779*. Paris, 1794.
———. *Voyages . . . au Japon, par le cap de Bonne-Espérance, les îles de la Sonde, &c.* 2nd ed. 4 vols. Paris, 1796.
Todd, T. Wingate, Beatrice Blackwood, and Harry Beecher. "Skin Pigmentation: The Color Top Method of Recording." *American Journal of Physical Anthropology* 11 (1928): 187–204.
Todd, T. Wingate, and Leona Van Gorder. "The Quantitative Determination of Black Pigmentation in the Skin of the American Negro." *American Journal of Physical Anthropology* 4 (1921): 239–60.
Toldt, Karl. "Über Hautzeichnung bei dichtbehaarten Säugetieren, insbesondere bei Primaten, nebst Bemerkungen über die Oberflächenprofilierung der Säugetierhaut." *Zoologische Jahrbücher. Abteilung für Systematik, Geographie und Biologie der Tiere* 35 (1913): 271–350.
Topinard, Paul. *L'anthropologie*. Paris: Reinwald, 1876.
———. *Anthropology*. Trans. Robert T. H. Bartley. London: Chapman and Hall, 1878.
———. *Éléments d'anthropologie générale*. Paris: Delahaye and Lecrosnier, 1885.
Torén, Olof. *En Ostindisk resa til Suratte, China, &c.* See Osbeck, Pehr.

———. *Voyage de Mons. Olof Torée*. Milan, 1771.
Torsellino, Orazio. *De vita S. Francisci Xaverii* (1594). Bologna, 1746.
Toynbee, Paget. *A Dictionary of Proper Names and Notable Matters in the Works of Dante*. Ed. Charles S. Singleton. Oxford: Clarendon Press, 1968.
Tredgold, A. F. *A Text-Book of Mental Deficiency (Amentia)*. 7th ed. London: Baillière, 1947.
Trigault, Nicolas. *See* Ricci, Matteo.
Tsu, Jing. *Failure, Nationalism, and Literature: The Making of Modern Chinese Identity, 1895–1937*. Stanford: Stanford University Press, 2005.
Tumpeer, I. Harrison. "Mongolian Idiocy in a Chinese Boy." *Journal of the American Medical Association* 79 (1922): 14–16.
Turner, Daniel. *De morbis cutaneis: A Treatise of Diseases Incident to the Skin*. London, 1714.
Turner, Samuel. *An Account of an Embassy to the Court of the Teshoo Lama, in Tibet*. London, 1800.
Tylor, Edward B. *Anthropology: An Introduction to the Study of Man and Civilization* (1881). London: Macmillan, 1924.
Valentijn, François. *Oud en nieuw Oost-Indien*. 5 vols. Dordrecht, 1724–26.
Valignano, Alessandro. *Historia del principio y progresso de la Compañía de Jesús en las Indias Orientales (1542–64)*. Ed. Josef Wicki. Rome: Institutum Historicum, 1944.
———. *Sumario de la cosas de Japón (1583)*. Ed. José Luis Alvarez-Taladriz. Tokyo: Sophia University, 1954.
Vallois, Henri Victor. "L'anthropologie physique." *Ethnologie générale*. Ed. Jean Poirier. Paris: Gallimard, 1968. 596–730.
Van Amringe, William Frederick. *An Investigation of the Theories of the Natural History of Man, by Lawrence, Prichard, and Others*. New York, 1848.
Van Braam Houckgeest, André Everard. *An Authentic Account of the Embassy of the Dutch East-India Company to the Court of the Emperor of China in the Years 1794 and 1795*. 2 vols. London, 1798.
———. *Voyage de l'ambassade de la Compagnie des Indes Orientales hollandaises vers l'empereur de la Chine dans les années 1794 & 1795*. 2 vols. Philadelphia, 1797–98.
Vaughn, Alden T. *Roots of American Racism: Essays on the Colonial Experience*. Oxford: Oxford University Press, 1995.
Virchow, Hans. "Von Luschan'schen Farbentafel zur Bestimmung der Hautfarbe." *Zeitschrift für Ethnologie* 58 (1926): 328.
Virey, Julien-Joseph. *Histoire naturelle du genre humain*. New ed. 3 vols. Paris, 1824.
Vogt, Carl. *Lectures on Man: His Place in Creation and in the History of the Earth*. Trans. James Hunt. London: Longman, 1864.
———. *Vorlesungen über den Menschen: seine Stellung in der Schöpfung und in der Geschichte der Erde*. 2 vols. Giessen: Ricker, 1863.
Voltaire. *Traité de métaphysique* (1734). Ed. H. Temple Patterson. Manchester: Manchester University Press, 1937.

Wagner, Henry R. "Mongolism in Orientals." *American Journal of Diseases of Children* 103 (1962): 706–14.
Walbaum, Christian Friedrich. *Ausführliche und merkwürdige Historie der Ost-Indischen Insel Groß-Java und aller übrigen holländischen Colonien in Ost-Indien.* Leipzig, 1754.
Wardle, Harriet Newell. "Evanescent Congenital Pigmentation in the Sacro-Lumbar Region." *American Anthropologist* 4 (1902): 412–40.
Washington, George. *The Papers of George Washington: Confederation Series.* Ed. W. W. Abbot and Dorothy Twohig. 10 vols. Charlottesville: University Press of Virginia, 1992–.
———. *The Writings of George Washington.* Ed. John C. Fitzpatrick. 39 vols. Washington, DC: Government Printing Office, 1931–44.
Weeks, Kent R. *The Valley of the Kings.* New York: Friedman/Fairfax, 2001.
Weiner, J. S. "A Spectrophotometer for Measurement of Skin Colour." *Man* 51 (1951): 152–53.
Weiner, Michael. "The Invention of Identity: Race and Nation in Pre-War Japan." *The Construction of Racial Identities in China and Japan: Historical and Contemporary Perspectives.* Ed. Frank Dikötter. Honolulu: University of Hawai'i Press, 1997. 96–117.
Weininger, Otto. *Geschlecht und Charakter: eine prinzipielle Untersuchung.* Vienna: Braumüller, 1903.
Wells, H. G. *The Outline of History, Being a Plain History of Life and Mankind.* 2 vols. New York: Macmillan, 1920.
———. *The Outline of History, Being a Plain History of Life and Mankind.* Rev. ed. 2 vols. New York: Macmillan, 1925.
Wernhart, Karl R. *Christoph Carl Fernberger: der erste österreichische Weltreisende (1621–1628).* Vienna: Europäischer Verlag, 1972.
What is Race? Evidence from Scientists. Paris: UNESCO, 1952.
Wilhelm II. *Briefe Wilhelms II an den Zaren, 1894–1914.* Ed. Walter Goetz. Berlin: Ullstein, 1920.
Williams, S. Wells. *The Middle Kingdom.* 2 vols. New York, 1848.
Winchell, Alexander. *Preadamites; or, a Demonstration of the Existence of Men Before Adam.* Chicago: Griggs, 1880.
Winckelmann, Johann Joachim. *The History of Ancient Art* (1764). Trans. G. Henry Lodge. 2 vols. Boston: Osgood, 1880.
Wippich, Rolf Harald. "Japan-Enthusiasm in Wilhelmine Germany: The Case of the Sino-Japanese War, 1894–5." *Japanese-German Relations, 1895–1945: War, Diplomacy, and Public Opinion.* Ed. Christian W. Sprang and Wippich. London: Routledge, 2006. 61–79.
Witsen, Nicolaas. *Noord en Oost Tartarye.* 2nd ed. Amsterdam, 1705.
Wittkower, Rudolf. *Allegory and the Migration of Symbols.* London: Thames and Hudson, 1977.

Wood, Anthony à. *Life and Times.* Ed. Andrew Clark. 5 vols. Oxford: Clarendon Press, 1891–1900.

Woodcock, O. H. "The History of the Medical Society of Individual Psychology." *Early Phases of Medical Psychology* (Individual Psychology Pamphlets, no. 23). London: Daniel, 1943. 11–16.

Wooler, T. J. *The Black Dwarf.* 12 vols. London, 1817–24.

Wreszinski, Walter. *Atlas zur altägyptischen Kulturgeschichte.* 2 vols. Leipzig: Hinrich, 1923–36.

Wright, Sewall. "The Effects in Combination of the Major Color-Factors of the Guinea Pig." *Genetics* 12 (1927): 530–69.

Wu, Frank H. *Yellow: Race in America Beyond Black and White.* New York: Basic Books, 2002.

Wu, William F. *The Yellow Peril: Chinese Americans in American Fiction, 1850–1940.* Hamden: Archon, 1982.

Xavier, St. Francis. *Epistolae.* Ed. Georg Schurhammer and Joseph Wicki. 2 vols. Rome: Monumenta Historica Societatis Jesu, 1944–45.

Young, John D. *Confucianism and Christianity: The First Encounter.* Hong Kong: Hong Kong University Press, 1983.

Young, Robert J. C. *Colonial Desire: Hybridity in Theory, Culture, and Race.* London: Routledge, 1995.

Yule, G. Udny. "On the Influence of Bias and of Personal Equation in Statistics of Ill-Defined Qualities." *Journal of the Anthropological Institute of Great Britain and Ireland* 36 (1906): 325–81.

Yule, Henry. *Cathay and the Way Thither: Being a Collection of Medieval Notices of China.* Rev. ed. Ed. Henri Cordier. 4 vols. London: Hakluyt Society, 1913–16.

Zammito, John H. "Policing Polygeneticism in Germany, 1775: (Kames,) Kant, and Blumenbach." *The German Invention of Race.* Ed. Sara Eigen and Mark Larrimore. Albany: State University of New York Press, 2006. 35–54.

Zarncke, Friedrich. "Der Priester Johannes." *Abhandlungen der philologisch-historischen Classe der Königlich Sächsischen Gesellschaft der Wissenschaften* 7 (1876): 829–1030; 8 (1879): 3–186 [repr. Hildesheim: Georg Olms, 1980].

Zhao, Rugua. *Chau Ju-Kua: His Work on the Chinese and Arab Trade in the Twelfth and Thirteenth Centuries, Entitled Chu-fan-chi.* Trans. Friedrich Hirth and W. W. Rockhill. St. Petersburg: Imperial Academy of Sciences, 1911.

———. *Zhufan zhi.* Taipei: Guangwen shuju, 1969.

Zihni, Lilian. "The History of the Relationship Between the Concept and Treatment of People with Down's Syndrome in Britain and America from 1866 to 1967." Ph.D. diss., University of London, 1989.

———. "Imitativeness and Down's Syndrome." *History and Philosophy of Psychology Newsletter* 19 (1994): 10–17.

Zimmermann, E.A.W. *Geographische Geschichte des Menschen und der allgemein verbreiteten vierfüßigen Thiere.* 3 vols. Leipzig, 1778–83.

Index

"active" or "passive" races, 80
Adachi, Buntaro, 110–11
African Americans, perceived as crossing between "true" Africans and generations of white and Native American interbreeding, 96
Ainu, 109, 136
Alvares, Jorge, 29
American Journal of Physical Anthropology, 88
Amiot, Joseph-Marie, 132–33
Ammon, Friedrich August von, varieties of epicanthi, 106
Annales du Musée Guimet, 19
Anthropological Institute of Great Britain, 21
Anthropological Society of London, 83
anthropology: correlation of cranial shape with intelligence, 71; criminal anthropology, 115; early concern with racial distinctions, 70–71; medical anthropology, 112–13; Mongolian race as yellow, 140–41; obsession with human measurement, 5, 70–71, 82–83; physical anthropology, 64, 72, 83, 84
anthropometry, 88
Ararat, Mount, 74
Aristotle, 43
"Aryan" race, 82
Ashmead, Albert, 111, 112
Asians: alteration of color from *fuscus* to *luridus*, 51–53; colorized without defined precedents, 44; distanced from whiteness embodied by West, 38–39; effeminized and objectified as potential sexual commodity, 48. *See also* East Asians
atavism, 111
Attila the Hun, 4, 75, 76, 143
Avicenna, 52

Bachman, John, 80, 81
Bacon, Francis, 35
Baelz, Erwin, 108–11, 123, 136
Barbosa, Duarte, 27, 102
Baur, Erwin, 113

"beauty" or "ugliness," as sign of human differences, 75
Beauvais, Vincent of. *See* Vincent of Beauvais
Beddoe, John: "index of nigrescence," 87
Belzoni, Giovanni Battista, 14; Procession of Egyptians, tomb of Seti I, color plate 1; Procession of "Ethiopians" and "Persians," tomb of Seti I, color plate 4; Procession of Jews, tomb of Seti I, color plate 2
Bernier, François, 3, 65, 103; description of yellow Indian women, 48; dual meaning of yellow skin, 52; "New Division of the Earth, According to the Different Species or Races of Man that Inhabit it, Sent by a Famous Voyager," 45–48, 50; reference to East Asians as *véritablement blanc* (truly white), 4
blackness, human: Chinese described as, 31; used to suggest sin and cultures outside Christian community, 23
black-white "crosses," 96
Blackwood, Beatrice, 98
Blumenbach, Johann Friedrich, 35, 44, 62, 65, 131; categorization of Asians, 41; Chinese and Japanese as subvarieties of Mongolian type, 65; on Chinese people, 66; concept of degeneration, 62; concept of "national faces" (*faciei gentilitiae*), 68; *De generis humani varietate nativa* (On the Natural Variety of Mankind), 59, 60, 61, 77; East Asians as antithetical to West, 123; East Asians as Mongolian, 4, 75; East Asians as true "dark" or "darkish olive" racial type, 61–62; East Asians as yellow or olive, 4, 61–62, 70; Egyptians as black, 17; five-race scheme, 42, 59–60, 64, 65, 114; five skulls from, 63; focus on exterior form only in classes of humans, 79; *Handbuch der Naturgeschichte* (Handbook of Natural History), 59, 60; invention of new racial category of Mongolian, 64, 70, 73, 125; measurement of human skulls, 5; popularization of term

Blumenbach, Johann Friedrich (*continued*) "Caucasian," 64; "racial face" of Mongolian type, 68–69; taxonomical schemes, 11; translation of Whang at Tong as "yellow man from the East," 67–68
"blushing race," Europeans as, 11
Bodin, Jean: *Six livres de la république*, 43–44
Boemus, Joannes: *Omnium gentium mores*, 44
The Book of Gates, 15
The Book of History, 136–37
Bory de Saint-Vincent, Jean Baptiste: *L'homme*, 17, 77, 78
Boxer Rebellion, 126
Bradley, Milton, 90, 96, 97, 155n12; *Elementary Color*, 91
Bradley, Richard: *Philosophical Account of the Works of Nature*, 50
British Association for the Advancement of Science, 84
Broca, Paul, 89; assessment of skin color, 5; chromatic table of eye colors, color plate 6; "Instructions générales pour les recherches et observations anthropologiques" (General Instructions for Anthropological Research and Observation), 83–84, 85, 86; "Tableau chromatique des yeux, de la peau et des cheveux pour les observations anthropologiques," 83–88, 93, color plate 6; table of hair and skin colors, color plate 7
Brosses, Charles de, 131–32
Brugsch, Henri: *Histoire d'Egypte*, 18
Bruno, Giordano, 45
Buddha-like position (cross-legged), 118
Buffon, George-Louis Leclerc, comte de, 44, 65; *Histoire naturelle, générale et particulière*, 58, 59; "Variétés dans l'espèce humaine, 58–59
Buryats, 73
Büttner, Christoph Gottlieb, 66

Camões, Luis de: *The Lusiads*, 25–26
Camper, Petrus, 76
Carpini, John of Plano: *See* John of Plano Carpini
Carus, Carl Gustav, 80
Catherine the Great, 73–74
"Caucasian," popularization of term, 64
Caucasoid, 123
Chambers, Robert: *Vestiges of the Natural History of Creation*, 105, 115, 122

Chambers, William: *Treatise of Oriental Gardening*, 67
Champollion, Jean-François: claim that ancient Egyptian racial categories were indistinguishable from nineteenth-century categories, 15–16; *Monuments de l'Égypte et de la Nubie*, 146n16
Champollion-Figeac, Jacques-Joseph: claim that Egyptians were Moorish, 17; "Peoples Known to the Egyptians," 18
China: first flag of Republic, 130; formulations associated with Five Elements, 133; human lightness and darkness as markers of perceived levels of civilization, 133; intractable to European religious and mercantile overtures, 30; as leading trader in Indian Ocean region, 25; reception of Western notion of yellow race, 8–9, 128–34, 137; reference to as "yellow terror" by 1870s, 125; Republican period, 129; yellow as color of earth and central color, 130; yellow cothing reserved for royal family, 35
Chinese, Western notions of: categorized as "yellow race," 42; children claimed to be white at birth and darkened as they grew older, 30; comparison to Germans, 27; darkening of in relation to difficulty of Christianizing, 30, 38, 42; described as brown, red, black, and swarthy by 1600, 31; described as white by early travelers, 147n11; not considered white, 36–37, 38; perception of sameness, 39–40; plethora of colors to describe, 30–34; view of as most permanent of races, 81
Chinese creation myth, 131
Chinese Exclusion Act of 1882, 138
Chinese immigrants, 126
Chinese physiognomic descriptions, 130–31
chinoiserie, eighteenth-century, 66, 76
climate, cited as factor in skin color, 17, 23, 29, 30, 31, 43, 59, 133, 147n5
color, measurement of: challenges of "yellow" and "red" skin, 85–86. *See also* color top
color charts, 6
color matching, 72
color table, 83–84, 87
color theory, 94
color top, 5, 6, 89–100, 123; ability to measure whiteness built into colors selected, 96; blackness of N disk, 95–96, 97; and

method of adjusting color disks, 93; seen as way of measuring blackness as opposed to other colors, 90
comparative anatomists, 5
Corsali, Andrea, 102
criminal anthropology, 115
Croix, Nicolle de la: *Géographie moderne*, 47, 62, 157n39
Crookshank, F. G.: *The Mongol in Our Midst: A Study of Man and His Three Faces*, 116–20; illustrations from, 119
Cruz, Gaspar da, 102, 140
Cuvier, Georges, 71, 78, 79, 82

Dante Alighieri: *Inferno*, three faces of Satan, 9
Darwin, Charles: *The Descent of Man*, 117; *The Origin of Species*, 83
Davenport, Charles, 90–91, 94; assumption that East Asian people were yellow, 98; biotypes of black African, 96; *Heredity in Relation to Eugenics*, 99; *The Heredity of Skin Color in Negro-White Crosses*, 92; *Inheritance in Poultry*, 91; measurement of aborigines and "black-white hybrids" in Australia, 96; rationalization for white disk of color top, 95; on skin pigmentation, 91–92; view that skin color of all races could be quantitatively described by color top, 97
"day" and "night" peoples, 80
degeneration, 62, 115, 122
Demel, Walter: "Wie die Chinesen gelb wurden" (How the Chinese Became Yellow), 3
Deniker, Joseph, 88–89
Desmoulins, Antoine, 78
Dikötter, Frank: *The Construction of Racial Identities in China and Japan*, 3; *The Discourse of Race in Modern China*, 155n10
Directions for Collecting Information and Specimens for Physical Anthropology, 88
Dixon, Roland, 140
Down, John Langdon, 113–16, 123
Down syndrome, 113–21
Du Halde, Jean-Baptiste, 73
Duméril, André, 78
Dutch East India Company, 103

East Asian bodies, in nineteenth-century medicine, 101–23; Mongolian eye, 101–7; Mongolian spots, 108–13

East Asian immigration, and need to racialize, 125
East Asians: described as *gilvus* (pale yellow), 62, 70; described as white in beginning of European exploration, 26–30; described as yellow, 1–2, 4, 9, 61–62, 70, 98; distinctions about skin color based on gender and social rank, 128; early European depictions of, 23–42; first depicted as yellow in art at beginning of nineteenth century, 9; perceived whiteness evaluative rather than descriptive, 28, 39; reception of color yellow, 128–34
Eden, Charles, 41–42
Eden, Richard, 35
Egyptians: debate about origins of, 16–18; perceived as "red," 14–15, 16, 18
Empoli, Giovanni da, 27
Encyclopaedia Britannica, 11th edition: 39, 135–36
Enlightenment, 82
epicanthus, 101, 103, 105, 106, 107, 114, 116, 118, 122, 123
epicanthus tarsalis, 107
Erxleben, Johann Christian Polykarp, 62, 65
Eskimos, 103, 110
Ethiopians, represented lower form of human life equivalent to infancy of other races, 106–7
ethnology, 71
eugenics, 90, 91
Europeans: appeal to natural phenomena to explain colors of people they encountered, 35; Asian peoples described as white in beginning of age of exploration, 26–30; considered as most civilized in taxonomic texts, 62; desire to find white Christians in Far East, 26; equation of human whiteness with Christianity, 25; as "the blushing race," 11
European world maps, tripartite form, 9–10
evolution, 83

facial angle, feature of taxonomies of racial difference, 17, 76, 81
Fernberger, Christoph Carl, 37
Fischer, Johann Eberhard: *Sibirische Geschichte* (Siberian History), 74, 113
"Five Elements" (*Wu xing*), 132
"Five Races Under One Union," 130

"four races of man," 6
"Fremdvölker-Expedition," 21
Fritsch, Gustav, 89
Fróis, Luis, 102
fuscus, as racial color term, 4, 49–51, 53–57, 62, 156n20

Gago, Balthasar, 28
Galen, 52
gender, and development of racialized thought, 48
Genesis, races of earth originated from sons of Noah, 12
Genghis Khan, 4, 75, 76, 143
Gentleman's Magazine, 67
Georgi, Johann Gottlieb: *Beschreibung aller Nationen des russischen Reichs* (Description of All Nations of the Russian Empire), 74
Gibbon, Edward, 125
Giddings, Franklin, 140; "Marking-Scheme for Nationalities," 142
Gliddon, George, 19, 81
Gmelin, Johann Friedrich, 57, 59
Gobineau, Arthur de: *Essai sur l'inégalité des races humaines* (Essay on the Inequality of the Human Races), 82, 125
Goethe, J. W. von: *Zur Farbenlehre,* 54
Goldsmith, Oliver: *History of the Earth,* 11–12, 62, 65
Gómara, Francisco López de: *Historia general de las Indias,* 34
González de Mendoza, Juan: *See* Mendoza, Juan González de
Gould, Stephen Jay, 114, 123
Greenland, 111–12
Grosier, Jean-Baptiste: *Description générale de la Chine,* 40
Gützlaff, Karl: *Sketch of Chinese History,* 40

Haddon, A. C.: *Races of Man,* 21, 139–40
Hakluyt, Richard: *Principal Navigations,* 73
Ham, sons of, 12, 80–81
Han people, 2, 129, 130
Herder, Johann Gottfried, and Mongol "barbarism," 76, 77
Herodotus, 44
Herskovits, Melville, 96
Hintze, Arthur, 89
Hippocrates, 52
Homo asiaticus, 4

homo sapiens, four colors of, 48–51
Hooton, Earnest, 140
Horn, Georg, 50
Hrdlicka, Ales, 88, 138–39
Huangdi, The Yellow Emperor, 67
human taxonomies, 11, 44, 62
Hume, David: "Of National Characters," 39–40
humoral theory, 44
Huns, 125
Huxley, Thomas, 115

Imperial Academy of Sciences, St. Petersburg, 74
India, use of "yellow" to describe people of, 45–48, 77
"Instructions générales pour les recherches et observations anthropologiques" (General Instructions for Anthropological Research and Observation), 83–84
Isidore of Seville: *Etymologies,* 10, 24, 147n5

Japan: alliance with Great Britain, 135; defeat of China and Russia, 7; end of seclusion policy, 41; foreign policy debates emphasizing Japanese identity, 135; reception of yellow designation, 9, 134–39; view of Yellow Peril as invention of jealous Western powers, 135; Western fears about aspirations as imperial power, 126; as "yellow hope," 138
Japanese: categorized as "yellow race," 41, 42; conversion to European Christianity, 29; described as brown or olive-skinned throughout eighteenth century, 41; lightened and darkened depending on Western preconceptions, 28, 42, 154n51; perceived as other than white by nonmissionary travelers, 29–30; persecution of Christians, 29; privileged over Chinese in Western discourse, 136; questions about racial makeup of West, 135–37; whiteness in missionary literature, 28–29, 148n13
Japhet, sons of, 12, 81
John of Plano Carpini, 73, 102
Jones, William, 67
Journal des sçavans, 45

Kaempfer, Engelbert: *History of Japan,* 29, 41, 103, 133, 136

Kalmucks, 73, 76, 131
Kang Youwei: *Da tongshu (Book of the Great Union)*, 129, 130
Kant, Immanuel, 74; "Bestimmung des Begriffs einer Menschenrace" (Determination of the Concept of a Human Race), 61; characterization of Indians as *gelb* or *olivengelb*, 47, 53, 61; characterization of Indians as "true" yellow, 4; color as infallible marker of racial difference, 61; division of world into four racial types, 60–61; "Von den verschiedenen Racen der Menschen" (On the Different Races of Man), 60
Klaatsch, Hermann, 117
Klemm, Gustav, 80
Kowner, Rotem, 3, 136
Kurz, Eugen, 117

"ladder approach" to race, 115
Latham, R. G., 80
Lawrence, James, 41–42
Lawrence, William, 78–79, 125
Leibniz, Gottfried Wilhelm, 45, 46
Lepsius, C. R.: *Denkmaeler aus Aegypten und Aethiopien*, 19, 146n16; figures from the tomb of Seti I, color plate 3
Li, Chi: *The Formation of the Chinese People*, 137
Lichtenberg, Georg Christoph, 67
Linnaeus, Carl, 44; all human beings composed of single species, 59; all of Asia considered "dark," 50; change in color of Asians to *braun*, 56–57; change of color of Asians from *fuscus* to *luridus*, 36, 41, 51–53, 56, 77; *Clavis medicinae*, 53; first to link yellow with Asia, 4; four colors of *homo sapiens*, 48–51; *Fundamenta botanica*, 53; interest in systematizing the world, 49; list of attributes for each human group, 54–55, 79; *Philosophia botanica*, 53; placement of Americans in separate group of peoples, 50; *Systema naturae*, 3–4, 48; table showing animal kingdom, 49; three of four varieties of *Homo sapiens*, 55
Linton, Ralph, 140
Liu, Chungshee H., 137
loess deposits, China, 139
Lombardi, Baldassare: racial reading of Satan's three faces, 10–12; vermillion face as representation of Europeans, 11; yellow-colored face as representation of Asians, 12
luridus: suggestion of ill health or disease, 52; used by Linnaeus to describe color of Asians, 36, 41, 51–53, 56, 77
Luschan, Felix von, 89, 93, 95, 97, 137

MacFarlane, Charles: *Japan*, 41
Maffei, Giovanni Pietro: *Historiarum indicarum*, 35
Malays, 5, 61, 62, 65, 77, 81, 103
Malte-Brun, Conrad, 77
Malthus, T. R., 76
Manchus, 73, 130
Mandeville, John: *Travels*, 52
Manuel I, 26, 27
Martini, Martino, 31, 102, 131; *De bello tartarico*, 73
Maxwell, James Clerk, 90
Maxwell disks, 90
medical anthropology, 112–13
medical "Mongolianness," 123
Meiners, Christoph, 82; designation of all East Asians as Mongolian, 75; division of humanity into two races, 74; *Grundriß der Geschichte der Menschheit* (Outline of the History of Man), 74; human differences centered on "beauty" or "ugliness," 75; Tartar peoples placed in "Caucasian" family, 74–75
Mellin, G. S.: *Encyclopädisches Wörterbuch der kritischen Philosophie* (Encyclopedic Dictionary of Critical Philosophy), 78
Mémoires concernant l'histoire, les sciences, les arts, les moeurs, les usages, &c. des Chinois: 131, 132
Mendelian laws of heredity, 90, 91, 96
Mendoza, Juan González de: *Historia . . . del gran reyno de la China*, 31–34, 38, 39, 73, 102; distinction between color of southern and northern residents, 50; "Of the temperature of the kingdome of China," 33
mental variations, quantified as racialized regressions from Caucasians, 115
Metchnikov, Elie, 107
Milton, John: *Paradise Lost*, 10
Minutoli, Johann Heinrich Carl, Freiherr von, 18; figures from the tomb of Seti I, color plate 5

missionaries, 1–2, 27–31, 39, 40, 42, 54, 75, 102, 128, 148n13
Moges, marquis de, 41, 42
Mongolenfleck, 109
Mongolia, protest to World Health Organization regarding use of "Mongolism," 120
Mongolian bodies, 121–23
Mongolian eye, 6, 101–7, 108, 111, 115, 121, 122
Mongolian fold, 107
"Mongolian permanence of type," 81–82
Mongolian race: yellow applied to, 60–65
Mongolian race, notion of, 60–65, 62; association with childishness, subhumanity, and underdevelopment, 7; characterization as dangerous, 4, 76, 99–100; continued use of term after World War II, 141; described as variety of colors, 77; distancing from white Western norm, 6; invention of in eighteenth century, 9, 43–69, 74; medical explanations for reinforced stereotypes of, 7; as "orangoid," 117; quantification of bodies, 6; quantification of skin color, 6; seen as "intermediate" race between white and black, 78; shift from position as "intermediate" variety to racial opposite, 64; as symbol of monstrosity, 76; viewed as invading race, 121; viewed as key feature of backward Far East, 76; viewed as nomadic and barbarous, 77, 79; *vs.* Tartar, 73–77
Mongolian spots, 6, 101, 108–13, 115, 118, 121, 122
Mongolism (now Down syndrome), 6, 101, 113–21; debates over terminology, 120–21
Mongoloid, 123
Mongols, 73, 125; as Golden Horde, 169n13; presumed characteristics, 75–76
Montagu, Ashley: commentary on Mongoloids, 141–42; *Man's Most Dangerous Myth*, 141–42
Montanus, Arnoldus, 35
Montesquieu, Charles de Secondat, baron de: *Lettres persanes*, 47
Morton, Samuel George: *Crania Aegyptiaca*, 17–18, 81, 82–89
Mungello, David: *The Great Encounter of China and the West*, 3
Münster, Sebastian: *Cosmographia*, 44, 102

Musa, Mark, 12
Myres, John L., 139

national faces, theory of, 5
Native Americans: categorization as yellow, 45; described as white by early travelers, 34; gradual shift to red, 1, 34, 44, 57, 59
natural science: growing confidence in biology and heredity, 71; notions of permanent hierarchies among races, 79. *See also* anthropology
natural selection, 83
Nazi theory, 75, 82
Negroid, 123
Nieuhof, Johannes, 102
Noah, 80
Noort, Olivier Van, 29
Notes and Queries on Anthropology for the Use of Travellers and Residents in Uncivilized Lands, 84, 88
Nott, Josiah, 19, 81
Nuwa, 131

"Observations on an Ethnic Classification of Idiots," 114
Odorico, da Pordenone, 24
Oken, Lorenz, 72
one-drop rule, 96
"orangoid" man, 117
Osbeck, Pehr, 56

Pallas, Peter Simon: *Sammlungen historischer Nachrichten über die mongolischen Völkerschaften* (Collection of Historical Information on the Mongolian Peoples), 74
Pan, Lynn: *Sons of the Yellow Emperor*, 2–3
Pantoja, Diego de, 39
Parrenin, Dominique, 39, 76
Parry, William Edward, 103
Parthians, 125
Pauthier, Guillaume: *Chine*, 81
Pearson, Karl, 92
Penrose, Lionel, 118
Perry, Matthew, 41
Persians, 125
phrenology, 71
physical anthropology, 64, 72, 83, 84
physiology, 71
Pires, Tomé: *Suma oriental*, 27
Pliny the Elder, 44, 131

Polo, Marco, 23–24, 44, 65, 73
polygenesis, 71, 72, 81, 83, 117
polyphyleticism, 117
Portuguese: emphasis on "whiteness" of East Asian people, 24; permanent outpost for East Asian trade at Malacca, 24
Pownall, Thomas, 62
Prester John, 24
Prichard, James Cowles, 42, 139; *Researches into the Physical History of Man,* 80
Psalmanazar, George, 40–41, 59

Qing Dynasty, 73
Quatrefages, Armand de, 125

racapitulation, 122
race: "active" or "passive," 80; anthropological concern with, 70–71; "Aryan," 82; distinctions viewed as permanent, 82; "ladder approach" to, 115; shift in thinking about during eighteenth century, 3; view of as permanent and biologically determined, 71. *See also* Western racial paradigms
race jaune, 77, 125
racial stereotypes, tools used to support notions of, 6
Rameses III, tomb of, 15, 19
Ramusio, G. B., 24, 26, 27
recapitulation, theory of, 105–6, 118, 123
red Egyptians, 14–15, 16, 18, 19, 21
red Native Americans, 1, 44, 57, 59
red skin: challenges to measures of, 85–86; and color top, 90, 94, 97; one of three faces of Satan, 9, 12; presumed characteristic of Chinese, 31, 32, 36; presumed of one of four "races of man," 6, 86
Rémusat, Abel, 78
retrogression, theories of, 121
Reynolds, Joshua, 66
Ricci, Matteo, 30, 50, 102, 131
Rogers, Spencer: *Colors of Mankind,* 21
Rosellini, Ippolito, 18, 146n16
Royal Anthropological Institute, 88
Rubruck, William of, 73, 102
Rudolphi, Karl, 77
Ryukyu islanders, 27

Sakyamuni Buddha, yellowness of face and robe, 131

Samoyeds, 85
Sánchez, Alonso, 31
Sande, Duarte de: *De missione legatorum Japonensium,* 50
Sapir, Edward, 118
Saris, John, 35
Satan: notion of three faces of read as reference to race, 9–13, 15; perceived as only truly black individual, 23
Sayers, Dorothy, 12
scientific racism, 59, 113
Scythians, 125, 131
Sequeira, Diogo Lopes de, 26
"Seres" and "Serica," 131
Sernigi, Girolamo, 24–25, 26
Serres, Etienne, 106–7
Seti I, tomb paintings, 45; claims that paintings showed yellow Asians, 19–21; nineteenth-century reproductions of, 18–21; paintings touted as representations of human ethnicity, 13–21, 133
Shem, 12
Shuttleworth, G. E., 116
Siebold, Philipp Franz von: *Nippon,* 41, 42, 103–5
Sino-Japanese War, 7, 126, 135
sinophilia, 76
sinophobia, 36
skin color: climatological theories of, 17, 23, 29, 30, 31, 43, 59, 133, 147n5
skin color, measurement of, 5, 10, 82–89; assumption of red, yellow, and black skin, 95
skull measurement, 81
skull shape, correlated with levels of intelligence, 71
Smith, Charles Hamilton, 80
Société d'anthropologie, Paris, 83, 84
Solinus, C. Julius, 147n5
"Sons of the Yellow Emperor," 2, 129, 133, 137
South Pacific peoples, 60
spectrophotometry, 98–99
Strabo, 131
Sullivan, Louis R.: *Essentials of Anthropometry,* 93
Sun Yat-sen: *Three Principles of the People,* 130, 138
survival of the fittest, 83

Taguchi, Ukichi, 135
Taiwan, 135

Tamerlane, 4, 75
Tan Chitqua (Tan Chetqua), 67
Tanegashima, 29
Tartar peoples, 65, 73, 125, 131
Tartary, 73
Tavernier, Jean-Baptiste, 76
"tawny moor," 45
Ternstroem, Christopher, 56
Thompson, Richard Austin, 126
three-ape thesis, 117
Thunberg, Carl Peter, 41, 42, 103
Tibetans, 73
Tilghman, Tench, 36
Toldt, Karl: examples of "Mongolian spots," 109
T-O maps, 9–10
Topinard, Paul: *Éléments d'anthropologie générale*, 21, 88, 95, 133
Torén, Olof, 56
Torsellino, Orazio, 30
travel narrators, perception of people of China and Japan as white, 23–24, 147n11
Trigault, Nicolas, 102
Turks, 125
Tylor, Edward, 11; *Anthropology*, 87
Types of Mankind (Nott and Gliddon): "The ancient Egyptian division of mankind into four species," 19–21, 81

UNESCO: "The Race Question," 141; *What is Race?*, 142
United States, immigration restrictions against Chinese and Japanese, 138

Valignano, Alessandro, 29, 30, 38
Valley of the Kings, 13
Vallois, Henri Victor, 142–43
Vasco de Gama, 24–25
Vincent of Beauvais: *Speculum maius*, 73
Virey, J.-J., 17, 77, 78
Vogt, Carl, 117
Voltaire: *Traité de métaphysique*, 47

Washington, George, 36
Western racial paradigms: desire to find yellowness in East Asian people, 72; imported into Chinese and Japanese contexts, 2, 7, 8–9; and need to distance whiteness from all others, 38–39, 72; no classification of people by race before end of eighteenth century, 23; notions about "inferior nature" of others, 78–79; obsession with chinoiserie, 76; preoccupation with determining color for each geographical group, 128
Whang at Tong, 66
"white Indians" of Central and South America, 96
whiteness: described as absence of color, 62, 72; description of in East Asians, 26–30; equation of with Christianity, 25; evaluative rather than descriptive for European explorers, 28; reserved for Western Europeans, 38–39
"white peril," 130
Wilhelm II, 7, 100, 138; *The Yellow Peril*, 126–28, 138
Williams, S. Wells, 125
Winchell, Alexander: *Preadamites*, 21
Winckelmann, Johann Joachim: *History of Ancient Art*, 17
Wreszinski, Walter: *Atlas zur altägyptischen Kulturgeschichte*, 21
Wu, Frank: *Yellow: Race in America Beyond Black and White*, 3

Xavier, St. Francis: "Letter to Europe," 28, 30, 102

yellow: applied to ancient Egyptians, 13–21; applied to East Asians, 1–2, 4, 7, 9, 40, 62, 70, 72, 139–44; applied to Japanese, 41, 42; applied to Mongolian race, 140–41; applied to Native Americans, 45; applied to North Africans, 45; applied to people in India, 45–48, 77; as "developmental" color, 122–23; dual meaning of yellow skin, 52; "good" and "bad," 53–57; in humoral theory, 52; as intermediate shade between white and black, 4, 6, 35, 52, 62; positive or negative connotations depending on language in which expressed, 34; reception of in China, 128–34; reception of in Japan, 134–39; significations in Chinese culture, 2; as term of complexional distance, 34
Yellow Emperor (Huangdi), 2, 35, 129

"Yellow Emperor's Canon of Internal Medicine" (*Huangdi neijing*), 129
yellow jaundice, 52
yellow pencil, 139
yellow peril, 7, 36, 100, 124–44, 143–44; invention of in 1895, 125; viewed as invention of jealous Western powers by Japan, 135
Yellow River, 2, 35, 130
"yellow soil people," 131

"The Yellow War," 7, 126
Yuan (Mongol) Dynasty, 73

Zappulla, Elio, 12
Zhao, Rugua: *Zhufan zhi (Description of Barbarous Peoples)*, 133
Zheng He (Chung Ho), 25
Zimmerman, E.A.W., 62; *Geographische Geschichte des Menschen* (Geographical History of Man), 74